Advances in Digitalization and Machine Learning for Integrated Building-Transportation Energy Systems

Advances in Digitalization and Machine Learning for Integrated Building-Transportation Energy Systems

Edited by

YUEKUAN ZHOU

Sustainable Energy and Environment Thrust, Function Hub, The Hong Kong University of Science and Technology (Guangzhou), Nansha, Guangdong, P.R. China; Department of Mechanical and Aerospace Engineering, The Hong Kong University of Science and Technology, Clear Water Bay, Hong Kong SAR, P.R. China; HKUST Shenzhen-Hong Kong Collaborative Innovation Research Institute, Futian, Shenzhen, P.R. China; Division of Emerging Interdisciplinary Areas, The Hong Kong University of Science and Technology, Clear Water Bay, Hong Kong SAR, P.R. China

JINGLEI YANG

The Hong Kong University of Science and Technology, Clear Water Bay, Hong Kong, P.R. China

GUOQIANG ZHANG

Hunan University, Changsha, Hunan, P.R. China

PETER D. LUND

Aalto University, Espoo, Finland

ELSEVIER

Elsevier
Radarweg 29, PO Box 211, 1000 AE Amsterdam, Netherlands
The Boulevard, Langford Lane, Kidlington, Oxford OX5 1GB, United Kingdom
50 Hampshire Street, 5th Floor, Cambridge, MA 02139, United States

Notices
Knowledge and best practice in this field are constantly changing. As new research and experience broaden our understanding, changes in research methods, professional practices, or medical treatment may become necessary.

Practitioners and researchers must always rely on their own experience and knowledge in evaluating and using any information, methods, compounds, or experiments described herein. In using such information or methods they should be mindful of their own safety and the safety of others, including parties for whom they have a professional responsibility.

To the fullest extent of the law, neither the Publisher nor the authors, contributors, or editors, assume any liability for any injury and/or damage to persons or property as a matter of products liability, negligence or otherwise, or from any use or operation of any methods, products, instructions, or ideas contained in the material herein.

ISBN: 978-0-443-13177-6

For Information on all Elsevier publications
visit our website at https://www.elsevier.com/books-and-journals

Publisher: Jonathan Simpson
Acquisitions Editor: Rachel Pomery
Editorial Project Manager: Aleksandra Packowska
Production Project Manager: Erragounta Saibabu Rao
Cover Designer: Vicky Pearson Esser

Typeset by MPS Limited, Chennai, India

Working together
to grow libraries in
developing countries

www.elsevier.com • www.bookaid.org

Dedication

We dedicate this first edition of *Advances in Digitalization and Machine Learning for Integrated Building-Transportation Energy Systems* to the members of the global integrated building-transportation energy systems community who continue to seek the best science, interpret that science for the better good of people worldwide, and persist in digitalization and machine learning approaches to better integrated building-transportation energy systems. We also dedicate this edition to our own scientific mentors and colleagues who have persuaded and occasionally pushed us in the direction of the highest quality building-transportation energy systems science—no matter the cost.

Prof. Yuekuan Zhou

Contents

11. Application of Internet of Energy and digitalization in smart grid and sustainability   211

Yuekuan Zhou

12. Application of big data and cloud computing for the development of integrated smart building transportation energy systems   223

Yuekuan Zhou

13. Social and economic analysis of integrated building transportation energy system   239

Zhengxuan Liu, Ying Sun and Ruopeng Huang

List of contributors

Tingxuan Chen
Department of Electrical Engineering, Shanghai Jiao Tong University, Shanghai,
P.R. China

Zhaohui Dan
Sustainable Energy and Environment Thrust, Function Hub, The Hong Kong University
of Science and Technology (Guangzhou), Nansha, Guangdong, P.R. China

Bin Gao
Sustainable Energy and Environment Thrust, Function Hub, The Hong Kong University
of Science and Technology (Guangzhou), Nansha, Guangdong, P.R. China

Ruopeng Huang
School of Management Science and Real Estate, Chongqing University, Chongqing,
P.R. China

Zhengxuan Liu
Faculty of Architecture and the Built Environment, Delft University of Technology,
Delft, The Netherlands

Zhuoxin Lu
Department of Electrical Engineering, University of Shanghai for Science and
Technology, Shanghai, P.R. China

Deng Pan
Sustainable Energy and Environment Thrust, Function Hub, The Hong Kong University
of Science and Technology (Guangzhou), Nansha, Guangdong, P.R. China

Ying Sun
School of Environmental and Municipal Engineering, Qingdao University of Technology,
Qingdao, P.R. China

Xiaoyuan Xu
Department of Electrical Engineering, Shanghai Jiao Tong University, Shanghai,
P.R. China

Xiaojun Yu
Sustainable Energy and Environment Thrust, Function Hub, The Hong Kong University
of Science and Technology (Guangzhou), Nansha, Guangdong, P.R. China

Xiaohan Zhang
Sustainable Energy and Environment Thrust, Function Hub, The Hong Kong University
of Science and Technology (Guangzhou), Nansha, Guangdong, P.R. China

Siqian Zheng
Department of Architecture and Civil Engineering, City University of Hong Kong,
Kowloon, Hong Kong SAR, P.R. China

Lu Zhou
Sustainable Energy and Environment Thrust, Function Hub, The Hong Kong University of Science and Technology (Guangzhou), Nansha, Guangdong, P.R. China

Yuekuan Zhou
Sustainable Energy and Environment Thrust, Function Hub, The Hong Kong University of Science and Technology (Guangzhou), Nansha, Guangdong, P.R. China; Department of Mechanical and Aerospace Engineering, The Hong Kong University of Science and Technology, Clear Water Bay, Hong Kong SAR, P.R. China; HKUST Shenzhen-Hong Kong Collaborative Innovation Research Institute, Futian, Shenzhen, P.R. China; Division of Emerging Interdisciplinary Areas, The Hong Kong University of Science and Technology, Clear Water Bay, Hong Kong SAR, P.R. China

Foreword

Timely information regarding digitalization and machine learning for integrated building-transportation energy systems is critical to improving energy efficiency and the robustness and reliability of the integrated energy-sharing network. We are pleased to present the first edition of *Advances in Digitalization and Machine Learning for Integrated Building-Transportation Energy Systems*.

The first edition of the book is launched with the goal to provide readers with the most comprehensive and current information covering the broad fields of digitalization and machine learning for integrated building transportation energy systems. The authors of this edition are from different countries and research institutions, reflecting a "who's who" of digitalization and machine learning for integrated building-transportation energy systems.

We trust this book will be a valuable resource for researchers, Artificial Intelligence and energy professionals, engineers, educators, and advanced energy field students. We are excited to advance the discipline of digitalization and machine learning for integrated building transportation energy systems with this publication.

ZHOU Yuekuan

Yuekuan Zhou, editor of the book

Preface

As editors, we feel privileged to have been asked to edit the first edition of *Advances in Digitalization and Machine Learning for Integrated Building-Transportation Energy Systems*. The first edition provides reference sources for smart buildings and intelligent transportations with artificial intelligence (AI) and digitalization technology, together with an explosion of exciting new methodologies and an understanding of the role of digitalization and machine learning in integrated building-transportation energy systems. AI and digital twins have relevance for energy sectors. Specifically, research indicated that digitalization and machine learning for integrated building-transportation energy systems will improve the system's operational robustness. Furthermore, AI can also be applied in integrated building-transportation energy systems for security, reliability, and stability. This work will be valued as an academic text in advanced energy courses. Recognizing the important role of the periodic updates of the book among scientists and practitioners, as editors we have sought to maintain the long-standing tradition of identifying content-thought leaders to provide the most comprehensive and latest information in their fields in the chapters represented in this edition.

The first edition of this book is presented with 13 chapters, including full-color illustrations and other color-enhanced features. The provision of 13 chapters will enable readers to quickly identify the location of relevant materials. At the end of each chapter, the authors have identified key research gaps and needs for future studies. Chapters in this book provide the latest scientific knowledge on smart buildings and intelligent transportations with AI and digitalization technology and also discuss important cross-disciplinary topics, including building occupant behavior and vehicle-driving schedule, electrification and hydrogenation in integrated building-transportation systems, hybrid energy storages in buildings with AI, smart grid with energy digitalization, peer-to-peer energy trading with AI, blockchain technologies for secure and tamper-proof in energy trading, energy economy and robustness with mobile energy storage systems, and social and economic analysis of integrated building transportation energy systems.

We believe the authors have done an outstanding job in presenting the latest information in their respective fields. We hope this edition will be an essential resource in the advanced field of energy.

Yuekuan Zhou
Jinglei Yang
Guoqiang Zhang
Peter D. Lund

Acknowledgments

The development of a book with profound analysis and highly scientific current content involves a large commitment of time and effort from many people. First, we would like to thank the authors of the 13 chapters for their commitment to this volume of reference work and, most importantly, their dedication to presenting the best information on the current status of science in their respective fields. Second, this edition would not have come to fruition without the untiring work and guidance of the four editors of the book, Yuekuan Zhou, Jinglei Yang, Guoqiang Zhang, and Peter D. Lund. We appreciate their guidance very much. Third, the success of any endeavor of this magnitude requires the support of family, colleagues, and friends to whom we owe our gratitude for their forbearance during the many hours devoted to this work's development and production.

CHAPTER 1

Smart buildings and intelligent transportations with artificial intelligence and digitalization technology

Deng Pan[1] and Yuekuan Zhou[1,2,3,4]
[1]Sustainable Energy and Environment Thrust, Function Hub, The Hong Kong University of Science and Technology (Guangzhou), Nansha, Guangdong, P.R. China
[2]Department of Mechanical and Aerospace Engineering, The Hong Kong University of Science and Technology, Clear Water Bay, Hong Kong SAR, P.R. China
[3]HKUST Shenzhen-Hong Kong Collaborative Innovation Research Institute, Futian, Shenzhen, P.R. China
[4]Division of Emerging Interdisciplinary Areas, The Hong Kong University of Science and Technology, Clear Water Bay, Hong Kong SAR, P.R. China

1.1 Introduction

Due to low energy efficiency, buildings currently account for over 30% of global energy consumption and one-third of carbon emissions [1]. Buildings and transportation contribute to a significant proportion of total energy consumption. Achieving carbon neutrality and sustainability calls for smart buildings and intelligent transportation systems. Digitalization is now becoming a new paradigm in the architecture and transportation industries [2]. To achieve carbon peaking and carbon neutrality, smart buildings with energy integration under e-mobility frameworks play a significant role in renewable penetration and grid independence. Furthermore, artificial intelligence (AI) and digitalization technology can promote a high-efficiency transition in smart buildings and transportation. In both academic and industrial circles, the application of AI and digitalization technology is regarded as a competitive technique for smart buildings integrated with intelligent transportations. This chapter provides a holistic overview of smart buildings and intelligent transportations with AI and digitalization technology.

1.2 Smart buildings

Buildings play a key role in energy consumption in cities. To improve energy efficiency, smart buildings utilize digitalization technology [3],

Advances in Digitalization and Machine Learning for Integrated Building-Transportation Energy Systems
DOI: https://doi.org/10.1016/B978-0-443-13177-6.00003-5

which includes six basic characteristics: automation, multifunctionality, adaptability, interactivity, efficiency, and intelligence [4]. Early smart building research only focused on the automated operation of building services systems, such as heat pumps, chillers, lighting, escalators, and so on [5]. Nowadays, with the deployment of renewable energy supply onsite and the role change from energy consumer to energy prosumer, peer-to-peer (P2P) energy sharing has attracted widespread interest among occupants and building owners. The increasing demand for sustainable and green buildings has directed the industry toward the development of smart buildings [6]. Energy savings generally sacrifice thermal comfort conditions. Quantifying building energy performance through the development and utilization of key performance characteristics is essential to achieving smart building goals [4]. The learning and application ability of buildings is considered one of the key performance indicators of smart buildings, which can improve energy performance over time based on accumulated experience (e.g., training data) [7]. Many new technologies are used in smart buildings, which mainly focuses on HVAC (heating, ventilation, and air conditioning) systems, energy flexible buildings, and P2P energy sharing.

1.2.1 Heating, ventilation, and air conditioning systems

HVAC systems account for almost 31% of the energy consumption in buildings [8]. HVAC systems are key areas to improve energy efficiency and reduce carbon emissions of smart buildings. Most studies mainly focus on energy management [3,9] and smart controls for HVAC systems. In terms of control strategies for HVAC systems of smart buildings, advanced control strategies include model predictive control and adaptive control methods [10]. A new HVAC control system with a multistep predictive deep reinforcement learning algorithm for smart buildings is proposed to save energy costs of the HVAC system while ensuring occupants' thermal comfort [11]. A two-stage conditional value-at-risk model is developed to optimize the day-ahead dispatch of smart buildings with HVAC systems [3]. Li et al. [12] proposed an event-driven multiagent-distributed optimal control strategy based on cyber-physical systems for HVAC. Chen et al. [13] presented a transfer learning model to design the HVAC control system and natural ventilation in a new smart building, which can reduce costs in control system design and commissioning. Chaouch et al. [9] proposed a smart method with fuzzy logic and machine-to-machine communication to control HVAC systems in smart buildings. Results show that the energy management system (EMS) can save 16% of annual energy consumption on average.

1.2.2 Energy flexibility

The energy flexibility of smart buildings refers to their capability in managing onsite renewable energy and energy demand in accordance with climate conditions, building demand, and grid requirements [14−16]. A new hierarchical optimization framework is designed to increase the joint flexibility of smart building communities [17]. A smart building is designed so that each power user has a flexible contract power to decrease electricity costs [18]. Energy flexibility is designed by thermal photovoltaics integrated with thermal storage systems for smart buildings to achieve smart energy management [19].

More and more smart appliances are being utilized in smart buildings. Thus the power load of smart buildings is more complicated. To precisely adjust the building power loads, a regulation structure using the load control method is developed [20]. Smart appliances can be dispatched spontaneously and harmonized with the EMS in a smart building, and the usage patterns of some flexible appliances (such as washing machines) can be used to shift loads by adjusting the operating time [21]. A multiobjective mixed-integer model using the weighted addition method is proposed for smart home dispatch, which can reduce 28.76% of operational costs and 11.37% of carbon emissions [22].

1.2.3 Peer-to-peer energy sharing

From the perspective of energy, prosumers in buildings with an onsite renewable power supply can be self-sufficient [23]. The energy sector is transforming from centralized energy systems to decentralized energy systems, which promotes the generation of P2P energy-sharing modes [24]. Renewables are often unstable and produce much energy surplus, which can be shared within the smart community [25]. Blockchain is being utilized to stimulate P2P energy trading due to its transparency, security, and high efficiency [26]. P2P energy sharing plays a key role in the renewable penetration ratio and energy flexibility of smart buildings. P2P energy sharing will provide additional energy flexibility and maximize profits for smart buildings by optimizing energy usage. Trading algorithms have been applied in the P2P energy trading market of smart energy communities to improve economic performance. Park et al. [27] designed a novel P2P trading mechanism to guarantee both flexibility and stability of energy supply within a smart community. Alam et al. [28] evaluated the energy costs' impact of P2P energy sharing among smart homes and proposed a near-optimal optimization algorithm to eliminate the unfair distribution of costs. Qiu et al. [29] proposed federated reinforcement learning to improve the economic and environmental performance of smart buildings when combined with P2P energy/carbon trading systems. Results show that it can reduce 5.87% of total energy costs and 8.02% of carbon emissions cost. Cutsem et al. [30] proposed a decentralized framework to manage the energy sharing of intelligent buildings using blockchain technology, which can be utilized within 100 intelligent buildings for a district community. Zhou et al. [31] adopted a P2P smart community energy pool and a user-dominated demand side response to improve usage efficiency and reduce the energy costs of an intelligent community.

1.3 Intelligent transportations

Transportation, which plays a key role in human society, is also one of the main sources of CO_2 emissions. However, the decarbonization potential in transportation is promising. With the increase in urban population, transportations face significant challenges, such as traffic congestion, safety, accidents, pollution issues, and energy consumption [32,33]. To solve these issues, new energy vehicles and energy savings in transportation are significant to decarbonize. Intelligent transportation systems have been rapidly developing in recent years, especially in fuel energy savings and decarbonization. Intelligent transportation systems have become one of the most important ways of decreasing traffic congestion, carbon emissions, and energy consumption [34]. Conventional intelligent transportation systems, include smart traffic lights, smart junction management, and intelligent traffic paths [35]. A smart EMS based on the road power demand model is utilized to decrease fuel consumption for transportations [33]. Zhao et al. [36] investigated the impact of intelligent transportation on carbon emissions in China. Results show that a 1% growth in intelligent transportation in a province can reduce the carbon emissions of the local and neighboring provinces by 0.1572% and 0.3535%, respectively. A multiscale carbon emissions computing platform is proposed for intelligent transportation to provide green transportation guidelines [37]. The impact of smart transportation systems on energy savings and decarbonization is investigated. Cui et al. [38] investigated the impact of different pathway guidance strategies on carbon emissions of smart transportations. Results show that a smoother path and a strategy with traffic data of the overall path will contribute to the decarbonization of intelligent transportation. Smart transportation systems use AI and digitalization technology to improve energy efficiency and decrease traffic jams, accidents, and emissions. Dynamic pricing mechanisms have a significant effect on smart transportation systems. An inappropriate dynamic pricing mechanism may result in more serious traffic stagnation, environmental pollution, and larger energy consumption [39]. The results of energy conservation and decarbonization are becoming primary rationales for investments in smart transportation [40].

Studies of intelligent transportation are also focusing on reducing traffic congestion. Smart transportation systems can plan highly effective real-time routes to reduce traffic congestion though smart communications [41]. Intelligent transportation utilizes ways of accessibility and emergency response for traffic control to solve traffic problems [42]. Rawashdeh et al. [43] established a communication framework to deal with intrusion threats to smart sensors of intelligent transportation systems. Traffic prediction based on machine learning can help plan pathways and control traffic congestion [44].

The digitization of the intelligent transportation system calls for the security and reliability of identity management and authentication [45]. The management and sensing systems of smart transportation may suffer from cyber-attacks [46]. Shari et al. [47] proposed a secure data dissemination project based on blockchain for intelligent transportation to achieve the security needs of reliability, privacy, and liability together.

Moreover, vehicular middleware and heuristic methods are utilized for intelligent transportation systems to achieve high mobility in transport and low postponement of the system [48]. Millimeter wave technologies are used to transfer large amounts of data efficiently for smart transportation [49]. Trading travel credit schemes for traffic jams, income protection, and Pareto-improving methods for peak-hour transit demand management are used for demand management strategies for intelligent transportation [50]. Nanogenerators-based self-powered sensors may be used in place of many sensors of intelligent transportation, which can accelerate the process to achieve complete smart transportation [51]. Intelligent transportation planning helps to build the optimal path to reduce traffic jams based on information, techniques, and models [52]. Apps for intelligent transportation can guide traffic with AI methods. AI with big data is used to construct a smart intelligence system. Traffic jams and path optimization selection utilize machine learning algorithms for prediction [53]. A smart transport system for the intelligent internet of vehicular network traffic is presented with tree-based machine learning models [54].

1.4 Artificial intelligence and digitalization technology

With the rapid development of information technology, AI and digitalization technology have been extensively used in all professions and trades, bringing innovation for smart buildings and intelligent transportations [55]. AI is considered a vital point for economic growth worldwide. Most countries invest a great deal of funds to develop AI and digitalization technology for the advancement of human society. The application of AI and digitalization technology in the building industry brings a new era for the development of design and construction companies. AI has been widely utilized in building service systems, mainly in demand forecasting and intelligent controls. Machine learning is used for solar radiation prediction [56]. The reinforcement learning approach can be used for smart charging/discharging on PCM (phase change material) storage to achieve more energy savings [57]. Anooj et al. [58] proposed a machine learning approach with a diagnosis for a PCM thermal management system to forecast liquid fraction. Zhou et al. [59] used supervised machine learning and heuristic optimization algorithms to optimize the design and operation of PCM coupled with a renewable system.

Machine learning for suitable AI applications in BIM models and combined optimization of topological rule inference will be the future trends [60]. The adoption of deep learning (DL) against the background of digital twins can drive the development of intelligent cities [61]. To accomplish more complex jobs in the modern world, intelligent industrial products need to be established with greater flexibility and adaptability. A digital twin provides for flexibility and adaptability [62]. Virtual sensing technologies play important roles in smart services in lifecycle construction and digitalization [63]. Virtual sensors are used in the HVAC systems of smart buildings for energy consumption monitoring, and they have also been broadly utilized in intelligent transportations for driving behavior and tracking. AI is applied to detect and diagnose building services systems using a labeled time series [64].

AI and digital twins are meaningful for the energy industry, producing profound effects on sustainable energy and the environment. The utilization of AI in smart buildings is an effective way of improving environmental and energy performance. AI and digitalization provide new chances for the transition from the conventional energy industries and producing new technologies that can improve the economic resilience of the energy sector [65]. AI technologies (such as big data, IoT, blockchain, cloud data storage, machine learning, and DL) help control energy data centers and P2P energy sharing, which have been increasingly utilized in the performance prediction of nonlinear energy systems [66]. Electricity system operators and utilities are widely applying AI techniques to cover daily operations and onsite service operations, which can optimize the operation of the power networks and enhance the flexibility, reliability, and efficiency of the grid [67]. A 3D indicator is proposed for AI utilization in the energy industry in terms of maturity level, supervisory risks, and potential benefits [68]. AI-generated renewables can learn from bioinspired lessons and provide smart energy systems for the carbon neutrality transition [69].

AI and digitalization technology are mainly used to predict renewable generation (PVs and wind turbines), smart energy management, and decarbonization. However, extreme disasters and climate change lead to more difficulties for AI prediction. Heo et al. [70] proposed an explicable AI-based generative model to assess techno–economic–environmental performance for predictive models of offshore wind turbines connected to power grids. Results show that 4% of the total cost of power generation and 41% of carbon emissions can be decreased. AI and digitalization provide new methods of monitoring and predicting carbon emissions. An energy and carbon footprint model trained by AI is used to assess and predict carbon emissions for different occupant densities of office buildings [71]. Machine learning and DL are used in advanced healthcare systems to reduce carbon emissions [72]. AI, machine learning, and deep learning are used for building design and visualization, structural design and analysis, intelligent operation, and lifecycle carbon emission analysis of the architecture and construction industry 4.0 [73]. Safety and efficiency of construction will be improved with the increase in accuracy and intelligence achieved through the application of AI. Debrah et al. [74] comprehensively reviewed AI utilization in green buildings and presented its future study directions in green buildings using digital twins, blockchain, robotics and 4D printing, and legal and moral duties of AI. Regona et al. [75]

applied AI technology to develop directions and limitations of the building sector in Australia through social media data analysis. Results showed that operational speed and digitalization are the main trends of AI in Australia. Data security and capabilities are the premier AI limitations for the building sector in Australia. Oluleye et al. [76] proposed a holistic framework, combining the AI models and the utilization field of the building product lifecycle. Fouquet et al. [77] studied the structural change with the dual transition of decarbonization and digitalization in European economies from a larger historical perspective. Results indicate that energy transitions emerge to be noticeably slower than digitalization transitions.

Building energy systems will produce different faults during operation. AI technology can help identify and classify different PV faults to recover from faults and achieve safe, reliable, and stable operations quickly and accurately. Mellit et al. [78] investigated different machine learning and ensemble learning approaches to detect and classify complex faults of PV arrays. Results show that the accuracy of fault detection is within 99%, while accuracy of fault classification is within 81.73%. Li et al. [79] developed a hybrid model by using a deep belief network enhanced by extreme learning machines to detect the refrigerant undercharge fault of chillers, increasing accuracy, and robustness. Zhu et al. [80] proposed domain adaptation transfer learning to build a diagnostic model for the targeted chiller with improved accuracy. Zhang et al. [81] holistically reviewed DL approaches for HVAC fault detection and diagnosis (FDD). Results show that a 2D convolutional neural network in DL is most widely used for HVAC FDD.

1.5 Smart buildings and intelligent transportations with artificial intelligence and digitalization technology

Transition to carbon–neutrality calls for combinations of smart buildings, intelligent transportations with AI, and digitalization technology. Tighter integration between smart buildings and transportations will reduce both CO_2 emissions and their energy consumption. Peak load shifting can reduce annual operational costs through the combined operation of smart buildings, intelligent transportation, and intelligent grids [82]. Zhou et al. [83] developed a district energy-sharing network and smart energy system, including district buildings, battery storage, transportation, and grid interconnection.

AI and digitalization technology can enhance integration between smart buildings and transportations. By regulating the bidirectional power flow with AI and digitalization technology, the interaction of smart buildings and transportations can reduce the dependence on the utility grid for both building and transportation energy usage. Studies have focused on energy integrations and interactions between buildings and transportation with multiobjective optimization, including energy consumption, carbon emissions, annual economic performance, and annual matching capacity [57]. AI and digitalization technology can be applied to optimize energy control strategies for integrating smart buildings with intelligent transportations, such as transportation travel schedules [84]. As shown in Fig. 1.1, Mirakhorli et al. [85] developed a model of predictive control to reduce the energy operation costs of smart homes, smart transportation, and smart grids. AI-driven intelligent strategies can improve the energy flexibility of smart buildings and intelligent transportations. Liu et al. [86] proposed a P2P energy trading strategy with multiobjective optimization methods on building energy systems combined with hydrogen and battery vehicles. Results show that it can decrease 18.54% of net grid import, 1594.13 tons of carbon emissions, 8.31% of net power rate, and US$ 458.69k of lifespan net present value. Zhou et al. [87] designed an interactive system between buildings and transportation with smart management approaches to enhance battery relative capacity. Teng et al. [88] proposed a novel stochastic analytical framework to quantify the profits of intelligent electric automobiles and heat pumps on CO_2 emissions and the integration cost of renewables in Britain. Results showed that intelligent electric automobiles and heat pumps reduced carbon emissions through more efficient operations.

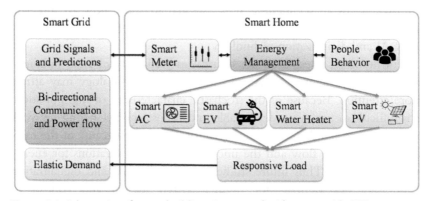

Figure 1.1 Schematics of smart buildings integrated with smart grids [85].

AI can be applied to smart buildings for intelligent transportations to enhance their robustness and adaptability to various uncertain scenarios. With the rapid development of intelligent sensor techniques, a large amount of data can be collected from smart buildings and intelligent transportations. Meanwhile, AI and digitalization technology are full of promising prospects in smart buildings and intelligent transportations in terms of robust design with scenario uncertainty, intelligent controls, FDD, and single and multiobjective optimizations [57]. However, there is a lack of studies on AI and digitalization technology in the context of the interaction between smart buildings and intelligent transportations.

1.6 Summary and future trends

This chapter provided a holistic introduction to smart buildings, intelligent transportations, intelligent and digitalization technology, and their combined interactions. A comprehensive literature review and in-depth discussion on smart buildings and intelligent transportations with AI and digitalization technology show that smart energy-sharing networks are promising to achieve carbon peaking and carbon neutrality. Both the smart building energy system and intelligent transportation can interact with each other through energy interaction mechanisms.

Regarding AI and digitalization technology in smart buildings and intelligent transportation, several key points need to be further considered before they can be widely accepted in the market. AI and digitalization technology in energy-sharing networks between smart buildings and intelligent transportations involve a large amount of cost for initial investment. Thus how to shorten the payback period of investment is important for owners' usage willingness. It is essential to study how to integrate intelligent transportation with smart building energy systems seamlessly using AI and digitalization technology.

Furthermore, research on the security and reliability of smart building-transportation systems with AI and digitalization technology may attract more attention. Because of the stochastic characteristics of transportation, it is difficult to achieve a dependable and robust energy-sharing network between smart buildings and transportation, especially because of the massive transportation flow and the uncertainty of time durations for energy interactions [82]. Thus future studies can focus on AI and digitalization technology development for optimal system design and control to enhance the robustness and reliability of the integrated energy-sharing network. To advance interactive energy-sharing networks of smart buildings and intelligent transportation, AI and digitalization technology need to be further investigated, including intelligent transportation scheduling optimization, layout optimization of smart charging stations and smart buildings, and policy incentives. Big data technology can be used for analyzing the distribution of transportation charging infrastructure and real driving data to help optimize its spatial distribution. Future research can also study the impact of uncertainty-based smart building demands and stochastic behaviors of intelligent transportation drivers on the techno-economic-environmental performance of the energy-sharing network.

References

[1] Casini M. Designing the third millennium's buildings. In: Smart Buildings 2016; p. 3−54
[2] Panteli C, Kylili A, Fokaides PA. Building information modelling applications in smart buildings: from design to commissioning and beyond a critical review. Journal of Cleaner Production 2020;265:121766.
[3] Feng W, Wei Z, Sun G, Zhou Y, Zang H, Chen S. A conditional value-at-risk-based dispatch approach for the energy management of smart buildings with HVAC systems. Electric Power Systems Research 2020;188:106535.
[4] Al Dakheel J, Del Pero C, Aste N, Leonforte F. Smart buildings features and key performance indicators: a review. Sustainable Cities and Society 2020;61:102328.
[5] Mofidi F, Akbari H. Intelligent buildings: an overview. Energy and Buildings 2020;223:110192.
[6] Omar O. Intelligent building, definitions, factors and evaluation criteria of selection. Alexandria Engineering Journal 2018;57:2903 -1.
[7] Alanne K. A novel performance indicator for the assessment of the learning ability of smart buildings. Sustainable Cities and Society 2021;72:103054.
[8] Kusiak A, Li M, Tang F. Modeling and optimization of HVAC energy consumption. Applied Energy 2010;87:3092−102.
[9] Chaouch H, Çeken C, Arı S. Energy management of HVAC systems in smart buildings by using fuzzy logic and M2M communication. Journal of Building Engineering 2021;44:102606.
[10] Gholamzadehmir M, Del Pero C, Buffa S, Fedrizzi R, Aste N. Adaptive-predictive control strategy for HVAC systems in smart buildings − a review. Sustainable Cities and Society 2020;63:102480.
[11] Liu X, Ren M, Yang Z, Yan G, Guo Y, Cheng L, et al. A multi-step predictive deep reinforcement learning algorithm for HVAC control systems in smart buildings. Energy. 2022;259:124857.
[12] Li W, Li H, Wang S. An event-driven multi-agent based distributed optimal control strategy for HVAC systems in IoT-enabled smart buildings. Automation in Construction 2021;132:103919.
[13] Chen Y, Tong Z, Zheng Y, Samuelson H, Norford L. Transfer learning with deep neural networks for model predictive control of HVAC and natural ventilation in smart buildings. Journal of Cleaner Production 2020;254:119866.
[14] Jensen SØ, Marszal-Pomianowska A, et al. IEA EBC annex 67 energy flexible buildings. Energy and Buildings 2017;155:25−34.

[15] Zhou Y, Cao S. Quantification of energy flexibility of residential net-zero-energy buildings involved with dynamic operations of hybrid energy storages and diversified energy conversion strategies. Sustainable Energy, Grids and Networks 2020;21:100304.

[16] Zhou Y, Cao S. Energy flexibility investigation of advanced grid-responsive energy control strategies with the static battery and electric vehicles: a case study of a high-rise office building in Hong Kong. Energy Conversion and Management 2019;199:111888.

[17] Angizeh F, Ghofrani A, Zaidan E, Jafari MA. Adaptable scheduling of smart building communities with thermal mapping and demand flexibility. Applied Energy 2022;310:118445.

[18] Foroozandeh Z, Ramos S, Soares J, Vale Z, Dias M. Single contract power optimization: a novel business model for smart buildings using intelligent energy management. International Journal of Electrical Power & Energy Systems 2022;135:107534.

[19] Maturo A, Buonomano A, Athienitis A. Design for energy flexibility in smart buildings through solar based and thermal storage systems: modelling, simulation and control for the system optimization. Energy 2022;260:125024.

[20] Zhang L, Tang Y, Zhou T, Tang C, Liang H, Zhang J. Research on flexible smart home appliance load participating in demand side response based on power direct control technology. Energy Reports 2022;8:424−34.

[21] Tang H, Wang S, Li H. Flexibility categorization, sources, capabilities and technologies for energy-flexible and grid-responsive buildings: state-of-the-art and future perspective. Energy 2021;219:119598.

[22] Zhen J, Khayatnezhad M. Optimum pricing of smart home appliances based on carbon emission and system cost. Energy Reports 2022;8:15027−39.

[23] Shamsuzzoha A, Nieminen J, Piya S, Rutledge K. Smart city for sustainable environment: a comparison of participatory strategies from Helsinki, Singapore and London. Cities (London, England) 2021;114:103194.

[24] Park BR, Chung MH, Moon JW. Becoming a building suitable for participation in peer-to-peer energy trading. Sustainable Cities and Society 2022;76:103436.

[25] Bandara KY, Thakur S, Breslin J. Flocking-based decentralised double auction for P2P energy trading within neighbourhoods. International Journal of Electrical Power & Energy Systems 2021;129:106766.

[26] Umar A, Kumar D, Ghose T. Blockchain-based decentralized energy intra-trading with battery storage flexibility in a community microgrid system. Applied Energy 2022;322:119544.

[27] Park S-W, Zhang Z, Li F, Son S-Y. Peer-to-peer trading-based efficient flexibility securing mechanism to support distribution system stability. Applied Energy 2021;285:116403.

[28] Alam MR, St-Hilaire M, Kunz T. Peer-to-peer energy trading among smart homes. Applied Energy 2019;238:1434−43.

[29] Qiu D, Xue J, Zhang T, Wang J, Sun M. Federated reinforcement learning for smart building joint peer-to-peer energy and carbon allowance trading. Applied Energy 2023;333:120526.

[30] Van Cutsem O, Ho Dac D, Boudou P, Kayal M. Cooperative energy management of a community of smart-buildings: a Blockchain approach. International Journal of Electrical Power & Energy Systems 2020;117:105643.

[31] Zhou S, Zou F, Wu Z, Gu W, Hong Q, Booth C. A smart community energy management scheme considering user dominated demand side response and P2P trading. International Journal of Electrical Power & Energy Systems 2020;114:105378.

[32] Sumalee A, Ho HW. Smarter and more connected: future intelligent transportation system. IATSS Research. 2018;42:67−71.

[33] Phan D, Bab-Hadiashar A, Lai CY, Crawford B, Hoseinnezhad R, Jazar RN, et al. Intelligent energy management system for conventional autonomous vehicles. Energy. 2020;191:116476.

[34] Yang Z, Peng J, Wu L, Ma C, Zou C, Wei N, et al. Speed-guided intelligent transportation system helps achieve low-carbon and green traffic: Evidence from real-world measurements. Journal of Cleaner Production 2020;268:122230.

[35] Kala R. Basics of intelligent transportation systems. In: On-Road Intelligent Vehicles; 2016; p. 401−19

[36] Zhao C, Wang K, Dong X, Dong K. Is smart transportation associated with reduced carbon emissions? The case of China. Energy Economics 2022;105:105715.

[37] Liu J, Li J, Chen Y, Lian S, Zeng J, Geng M, et al. Multi-scale urban passenger transportation CO_2 emission calculation platform for smart mobility management. Applied Energy 2023;331:120407.

[38] Cui N, Chen B, Zhang K, Zhang Y, Liu X, Zhou J. Effects of route guidance strategies on traffic emissions in intelligent transportation systems. Physica A: Statistical Mechanics and its Applications 2019;513:32−44.

[39] Saharan S, Bawa S, Kumar N. Dynamic pricing techniques for Intelligent Transportation System in smart cities: a systematic review. Computer Communications 2020;150:603−25.

[40] Chen Y, Ardila-Gomez A, Frame G. Achieving energy savings by intelligent transportation systems investments in the context of smart cities. Transportation Research Part D: Transport and Environment 2017;54:381−96.

[41] Regragui Y, Moussa N. A real-time path planning for reducing vehicles traveling time in cooperative-intelligent transportation systems. Simulation Modelling Practice and Theory 2023;123:102710.

[42] Almutairi S, Manimurugan S. S. Non-divergent traffic management scheme using classification learning for smart transportation systems. Computers and Electrical Engineering 2023;106:108581.

[43] Rawashdeh M, Alshboul Y, Zamil MGHAL, Samarah S, Alnusair A, Hossain MS. A security framework for QaaS model in intelligent transportation systems. Microprocessors and Microsystems 2022;90:104500.

[44] Boukerche A, Tao Y, Sun P. Artificial intelligence-based vehicular traffic flow prediction methods for supporting intelligent transportation systems. Computer Networks 2020;182:107484.

[45] Das D, Dasgupta K, Biswas U. A secure blockchain-enabled vehicle identity management framework for intelligent transportation systems. Computers and Electrical Engineering 2023;105:108535.

[46] Ganin AA, Mersky AC, Jin AS, Kitsak M, Keisler JM, Linkov I. Resilience in intelligent transportation systems (ITS). Transportation Research Part C: Emerging Technologies 2019;100:318−29.

[47] Mohd Shari NF, Malip A. Blockchain-based decentralized data dissemination scheme in smart transportation. Journal of Systems Architecture 2023;134:102800.

[48] Kumar R, Khanna R, Kumar S. Vehicular middleware and heuristic approaches for intelligent transportation system of smart cities. Cognitive Computing for Human-Robot Interaction 2021;163−75.

[49] Matrouk K, Trabelsi Y, Gomathy V, Arun Kumar U, Rathish CR, Parthasarathy P. Energy efficient data transmission in intelligent transportation system (ITS): millimeter (mm wave) based routing algorithm for connected vehicles. Optik. 2023;273:170374.

[50] Qin X, Ke J, Wang X, Tang Y, Yang H. Demand management for smart transportation: a review. Multimodal Transportation 2022;1:100038.

[51] Askari H, Khajepour A, Khamesee MB, Wang ZL. Embedded self-powered sensing systems for smart vehicles and intelligent transportation. Nano Energy 2019;66:104103.

[52] Karami Z, Kashef R. Smart transportation planning: Data, models, and algorithms. Transportation Engineering 2020;2:100013.

[53] Iyer LS. AI enabled applications towards intelligent transportation. Transportation Engineering 2021;5:100083.

[54] Rani P, Sharma R. Intelligent transportation system for internet of vehicles based vehicular networks for smart cities. Computers and Electrical Engineering 2023;105:108543.

[55] Wu J, Wang X, Dang Y, Lv Z. Digital twins and artificial intelligence in transportation infrastructure: classification, application, and future research directions. Computers and Electrical Engineering 2022;101:107983.

[56] Deo RC, Şahin M, Adamowski JF, Mi J. Universally deployable extreme learning machines integrated with remotely sensed MODIS satellite predictors over Australia to forecast global solar radiation: a new approach. Renewable and Sustainable Energy Reviews 2019;104:235−61.

[57] Zhou Y. Artificial intelligence in renewable systems for transformation towards intelligent buildings. Energy and AI 2022;10:100182.

[58] Anooj GVS, Marri GK, Balaji C. A machine learning methodology for the diagnosis of phase change material-based thermal management systems. Applied Thermal Engineering 2023;222:119864.

[59] Zhou Y, Zheng S, Zhang G. Machine learning-based optimal design of a phase change material integrated renewable system with on-site PV, radiative cooling and hybrid ventilations—study of modelling and application in five climatic regions. Energy 2020;192:116608.

[60] Sacks R, Girolami M, Brilakis I. Building information modelling, artificial intelligence and construction tech. Developments in the Built Environment 2020;4:100011.

[61] Wang W, He F, Li Y, Tang S, Li X, Xia J, et al. Data information processing of traffic digital twins in smart cities using edge intelligent federation learning. Information Processing & Management 2023;60:103171.

[62] Lin TY, Jia Z, Yang C, Xiao Y, Lan S, Shi G, et al. Evolutionary digital twin: a new approach for intelligent industrial product development. Advanced Engineering Informatics 2021;47:101209.

[63] Yoon S. Virtual sensing in intelligent buildings and digitalization. Automation in Construction 2022;143:104578.

[64] Xie X, Merino J, Moretti N, Pauwels P, Chang JY, Parlikad A. Digital twin enabled fault detection and diagnosis process for building HVAC systems. Automation in Construction 2023;146:104695.

[65] Lei Y, Liang Z, Ruan P. Evaluation on the impact of digital transformation on the economic resilience of the energy industry in the context of artificial intelligence. Energy Reports 2023;9:785−92.

[66] Lyu W, Liu J. Artificial Intelligence and emerging digital technologies in the energy sector. Applied Energy 2021;303:117615.

[67] Ahmad T, Zhang D, Huang C, Zhang H, Dai N, Song Y, et al. Artificial intelligence in sustainable energy industry: status quo, challenges and opportunities. Journal of Cleaner Production 2021;289:125834.

[68] Quest H, Cauz M, Heymann F, Rod C, Perret L, Ballif C, et al. A 3D indicator for guiding AI applications in the energy sector. Energy and AI 2022;9:100167.

[69] Liu Z, Sun Y, Xing C, Liu J, He Y, Zhou Y, et al. Artificial intelligence powered large-scale renewable integrations in multi-energy systems for carbon neutrality transition: challenges and future perspectives. Energy and AI 2022;10:100195.

[70] Heo S, Ko J, Kim S, Jeong C, Hwangbo S, Yoo C. Explainable AI-driven net-zero carbon roadmap for petrochemical industry considering stochastic scenarios of remotely sensed offshore wind energy. Journal of Cleaner Production 2022;379:134793.

[71] Chen C-Y, Chai KK, Lau E. AI-Assisted approach for building energy and carbon footprint modeling. Energy and AI 2021;5:100091.

[72] Das KP, Chandra J. A survey on artificial intelligence for reducing the climate footprint in healthcare. Energy Nexus 2023;9:100167.

[73] Baduge SK, Thilakarathna S, Perera JS, Arashpour M, Sharafi P, Teodosio B, et al. Artificial intelligence and smart vision for building and construction 4.0: machine and deep learning methods and applications. Automation in Construction 2022;141:104440.

[74] Debrah C, Chan APC, Darko A. Artificial intelligence in green building. Automation in Construction 2022;137:104192.

[75] Regona M, Yigitcanlar T, Xia B, Li RYM. Artificial intelligent technologies for the construction industry: how are they perceived and utilized in Australia? Journal of Open Innovation: Technology, Market, and Complexity 2022;8.

[76] Oluleye BI, Chan DWM, Antwi-Afari P. Adopting Artificial Intelligence for enhancing the implementation of systemic circularity in the construction industry: a critical review. Sustainable Production and Consumption 2023;35:509−24.

[77] Fouquet R, Hippe R. Twin transitions of decarbonisation and digitalisation: a historical perspective on energy and information in European economies. Energy Research & Social Science 2022;91:102736.

[78] Mellit A, Kalogirou S. Assessment of machine learning and ensemble methods for fault diagnosis of photovoltaic systems. Renewable Energy 2022;184:1074−90.

[79] Li P, Anduv B, Zhu X, Jin X, Du Z. Diagnosis for the refrigerant undercharge fault of chiller using deep belief network enhanced extreme learning machine. Sustainable Energy Technologies and Assessments 2023;55:102977.

[80] Zhu X, Chen K, Anduv B, Jin X, Du Z. Transfer learning based methodology for migration and application of fault detection and diagnosis between building chillers for improving energy efficiency. Building and Environment 2021;200:107957.

[81] Zhang F, Saeed N, Sadeghian P. Deep learning in fault detection and diagnosis of building HVAC systems: a systematic review with meta analysis. Energy and AI 2023;12:100235.

[82] Zhou Y, Cao S, Hensen JLM, Lund PD. Energy integration and interaction between buildings and vehicles: a state-of-the-art review. Renewable and Sustainable Energy Reviews 2019;114:109337.

[83] Zhou Y, Cao S, Hensen JLM. An energy paradigm transition framework from negative towards positive district energy sharing networks—battery cycling aging, advanced battery management strategies, flexible vehicles-to-buildings interactions, uncertainty and sensitivity analysis. Applied Energy 2021;288:116606.

[84] Amirioun MH, Kazemi A. A new model based on optimal scheduling of combined energy exchange modes for aggregation of electric vehicles in a residential complex. Energy 2014;69(5):186−98.

[85] Mirakhorli A, Dong B. Market and behavior driven predictive energy management for residential buildings. Sustainable Cities and Society 2018;38:723−35.

[86] Liu J, Yang H, Zhou Y. Peer-to-peer trading optimizations on net-zero energy communities with energy storage of hydrogen and battery vehicles. Applied Energy 2021;302:117578.

[87] Zhou Y, Cao S. Coordinated multi-criteria framework for cycling aging-based battery storage management strategies for positive building—vehicle system with renewable depreciation: life-cycle based techno-economic feasibility study. Energy Conversion and Management 2020;226:113473.

[88] Teng F, Aunedi M, Strbac G. Benefits of flexibility from smart electrified transportation and heating in the future UK electricity system. Applied Energy 2016;167:420−31.

CHAPTER 2

Machine learning and artificial intelligence-distributed renewable energy sources: technologies, perspectives, and challenges

Xiaojun Yu[1] and Yuekuan Zhou[1,2,3,4]

[1]Sustainable Energy and Environment Thrust, Function Hub, The Hong Kong University of Science and Technology (Guangzhou), Nansha, Guangdong, P.R. China
[2]Department of Mechanical and Aerospace Engineering, The Hong Kong University of Science and Technology, Clear Water Bay, Hong Kong SAR, P.R. China
[3]HKUST Shenzhen-Hong Kong Collaborative Innovation Research Institute, Futian, Shenzhen, P.R. China
[4]Division of Emerging Interdisciplinary Areas, The Hong Kong University of Science and Technology, Clear Water Bay, Hong Kong SAR, P.R. China

2.1 Introduction: artificial intelligence for smart energy systems

Deep learning (DL) and reinforcement learning (RL) show strong capabilities in process systems engineering and multistage decision [1]. Contrary to the traditional iteration process that follows physical laws, machine learning (ML) is capable of establishing a straightforward mathematical association between inputs and outputs. By significantly reducing computational time during times of iteration convergence for optimization and decision-making, ML shows promising potential in system engineering applications [2–4]. Applications of artificial intelligence (AI) in smart energy systems include performance forecasting [5–7]; adaptive controls, such as maximum power point tracking (MPPT) on solar photovoltaics (PVs) [8], passive stall control [9], and active pitch control [10] in wind turbines; and optimizations [11].

ML and AI can improve energy conversion and storage efficiency, thereby reducing carbon emissions. Through a comprehensive review of RL in energy management, 10%–20% of performance can be improved [12], while the increased complexity of future energy hubs requires continuous breakthroughs in computing technologies. ML in nanofluid heat

Advances in Digitalization and Machine Learning for Integrated
Building-Transportation Energy Systems
DOI: https://doi.org/10.1016/B978-0-443-13177-6.00012-6

transfer enhancement [13] can effectively improve solar PV efficiency. Human-building interaction [14] with DL will improve energy performance in buildings, and energy efficiency in building services systems. By establishing a relationship between renewable energy and coal and carbon emissions, Magazzino et al. [15] indicated that the increased penetration of solar-wind energy in national energy structures will reduce carbon emissions in China and the US, while stricter restrictions on coal consumption are necessary to decarbonize the energy systems in India. Additionally, DL in demand and energy supply can guide sustainable policy in Korea [16]. ML-based prediction on CO_2 emission intensity in the power grid [17] can provide guidelines on clean transformation. Sharma et al. [18] reviewed recent advances in various ML algorithms in nanofluid heat transfer enhancement in renewable systems and energy efficiency improvement. In terms of optimal operation on distributed renewable systems, Specht and Madlener [19] trained RL algorithms to manage small-scale assets.

In terms of technical investment on renewable systems (e.g., wind power, solar PV, hydropower, and biomass), ML can assist in investment decision-making, taking into consideration the GDP rate, inflation rate, electricity demand, electricity generation, electricity price, and CO_2 emissions [20–22]. Electricity generation and electricity price are dominating factors in deciding on renewable energy investment. The DL-driven multiscale smart energy system can assist in design, planning, and service [23].

This chapter aims to provide an overview of ML in the digital energy era with respect to performance prediction and optimization under deterministic and probabilistic scenarios. With supercomputing enabling accurate performance prediction of nonlinear systems, ML in model predictive control (MPC) has also been reviewed. The future prospects of ML lie in energy transition and sustainable development goals (SDGs).

2.2 Sustainable transition with cleaner power production and machine learning

Smart sensors/metering enable the energy digitalization transformation [24], while AI with supercomputing and data processing capability will increasingly play an inevitable role in sustainable transition. Moreover, through simulation, extension, and expansion of human intelligence, AI will enable digital energy systems to become intelligent and energy efficient. Frontier guidelines on AI applications were provided by Zhou [25] for intelligent buildings, involving distributed renewable integration, demand-response, and fault detection and diagnosis. This subsection aims to uncover the fundamental role of ML in multienergy systems.

Fig. 2.1 demonstrates ML-based optimization in nonlinear renewable systems. Initially, the experimental database is classified as a training and testing database. With respect to the training database, ML models will be well trained through probabilistic prediction and hyperparameter optimization. The testing database is to validate the model's accuracy. The well-trained model will be applied for sensitivity analysis and multivariate optimization when integrated with an objective function.

2.2.1 Machine learning for deterministic power prediction

With strong capabilities in inherent nonlinear features and high-level invariant structures, Wang et al. [28] reviewed DL in renewable energy forecasting using adopting deterministic and probabilistic forecasting methods. Sharifzadeh et al. [29] compared the prediction performance of integrated solar-wind energy systems among conventional artificial neural networks, support vector regression, and Gaussian process regression. Aslam et al. [30] comprehensively reviewed DL in power load and renewable energy forecasting. Results highlight the significance of a large number of data points to ensure the reliability of the performance prediction. Liu et al. [31] comprehensively reviewed AI for deterministic wind energy prediction. Ensemble learning and metaheuristic optimization have both been deployed as intelligent predictors. Table 2.1 summarizes ML-based predictions on various renewable systems. Robust prediction performance on various renewable systems with high prediction accuracy demonstrates the feasibility and flexibility of ML in dynamic energy performance prediction.

2.2.2 Machine learning for probabilistic performance prediction

In addition to deterministic power prediction [43], probabilistic performance prediction of renewable systems has also been conducted under uncertainty. Wang et al. [44] applied a Bayesian-based adaptive robust multikernel regression model to overcome the disadvantages of traditional models, that is, high-resolution data ignorance and inconsistency with real wind power under error distribution. Zhou et al. [43] trained a surrogate model to predict solar PV power under uncertainty. Results indicated that

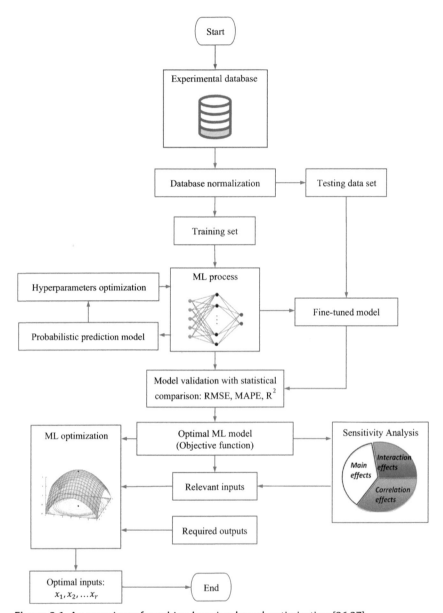

Figure 2.1 An overview of machine learning-based optimization [26,27].

probabilistic performance prediction is necessary to avoid performance underestimation. Mashlakov et al. [45] conducted probabilistic energy forecasting using DL models. Results indicated that the reliability of the model is dependent on dataset heterogeneity, hyperparameters, and

Table 2.1 Machine learning for predictions of various renewable systems.

Renewable systems	Studies	Machine learning algorithms	Results
Wind power	Lin and Liu [32]	Deep learning algorithm	Computational cost and time can be reduced with high accuracy
	Hong and Rioflorido [33]	Double Gaussian activation function	Coefficient of determination is higher than 0.88 all year round.
Solar energy	Mellit et al. [34]	Long short-term memory	Good accuracy with acceptable correlation coefficient between 96.9% and 98%
	Voyant et al. [35]	Regression tree, random forest, and gradient boosting	Hybrid models or ensemble forecasting approach can improve prediction accuracy.
	Wang et al. [36]	Convolutional neural network and long short-term memory	Hybrid models show the best performance in PV power prediction, followed by convolutional neural network.
	Qin et al. [37]	Data-driven model	Cloud fraction and solar zenith angle are two critical factors in model accuracy.
Hydrogen energy	Lee et al. [38]	Back-propagation algorithm	Thermal efficiency of 85.6% can be realized through hyperparameter optimization.
	Hwangbo et al. [39]	Deep learning algorithm	Deep learning-based supply-demand forecasting can sustain the hydrogen network with 64.5% total environmental cost savings.

(*Continued*)

Table 2.1 (Continued)

Renewable systems	Studies	Machine learning algorithms	Results
Biomass	Long et al. [40]	Support vector regression	Machine learning-assisted design of algal cultivation for sustainable biofuel production.
	Potnuri et al. [41]	Polynomial regression	Machine learning-based prediction on catalytic pyrolysis of biomass with $R^2 > 0.93$
	Naqvi et al. [42]	Deep learning	Thermochemical conversion of biomass can be accurately modeled by machine learning through classification, regression, and optimization.

hardware. Both deterministic and probabilistic forecasting approaches are studied to predict PV power [46], and the prediction reliability can be improved over conventional forecasting models.

Based on ML, Zhang et al. [47] conducted probabilistic interval prediction on wind power, showing reliable and excellent prediction results. In respect to wind energy, Xie et al. [48] reviewed deterministic and probabilistic forecasting methods for wind energy forecasting, with respect to data sources, model framework, and prediction performance assessment. Based on prediction results from both renewable energy and building demand, Shen et al. [49] developed a multiobjective cooperation optimization to achieve cooperative control over building energy systems. Results indicated that, compared with traditional rule-based control, the multiagent cooperation algorithm can improve the self-consumption of renewable energy by 43%, and reduce the energy cost by 8%. Harrold et al. [50] applied multiagent deep RL for renewable integration in microgrids, including peer-to-peer energy sharing and selling power back to the grid.

2.2.3 Machine learning for multivariate optimization under deterministic scenarios

The straightforward mathematical association between inputs and outputs with ML can accelerate the search for optimal solutions. Zhou and Zheng [6] conducted multivariate optimization on an aerogel glazing in different climates. By replacing the traditional physical model with the surrogate model, the optimization process was time saving. Furthermore, compared with the Taguchi orthogonal array, the optimal results can reduce the total heat gain by 62.5% in LanZhou and by 5.9% in subtropical Guangzhou. Zhang et al. [51] studied hierarchical decision-making in wind power management via deep RL. The model-free prediction can effectively reduce operating costs. By using the supervised learning approach for surrogate model training, Zhou et al. [52] optimized both the design and operating parameters on a solar PV system. Results showed enhanced solar power supply by timely cooling down the solar cell. Wang et al. [53] optimized solar-driven hydrogen production using an ML-based optimization approach. By simultaneously considering battery safety and off-grid operation, Gao et al. [54] applied deep RL for bi-objective optimization. The case study verifies the effectiveness of the proposed approach. Yi et al. [55] applied deep RL for optimal decision-making for nuclear-renewable systems to maximize revenue. Results verify the generalization capability of the proposed approach.

2.2.4 Machine learning for multivariate optimization under stochastic uncertainty

In addition to traditional optimization under deterministic scenarios, optimization under stochastic uncertainty is necessary, especially considering epistemic and aleatory uncertainties. The holistic overview and comparison of uncertainty-based optimization can contribute to the further development of multistage stochastic programs and uncertainty-based decision-making [56]. Zakaria et al. [57] reviewed uncertainty approaches for stochastic optimization. The review highlights the superiority of stochastic optimization over deterministic approaches. Guevara et al. [58] applied robust optimization in energy planning strategies, considering uncertainty. The role of ML is to restrict uncertain parameters and identify the important parameters to guarantee the efficiency of the model. Crespo-Vazquez et al. [59] conducted ML-based stochastic optimization for decision-making on the participation of wind and storage power in the energy market with uncertainty in energy prices. The approach can enable robust decisions with competitive economic benefits. Zhou and Zheng [60] developed an uncertainty optimization approach for multivariate optimization, considering input uncertainties. Results indicated that uncertainty optimization can further increase PV performance over the stochastic uncertainty scenario. Considering the stochastic uncertainty magnitude and sampling size, Zhou [61] conducted a two-stage learning approach to identify the uncertainty-based optimal case. Results indicate that the uncertainty-based optimal solution is highly dependent on uncertainty magnitudes, but less dependent on the stochastic sampling size.

2.3 Model predictive controls with machine learning in renewable energy sources

The main advantages of MPC include building dynamics, predictions on future disturbances, constraints, and conflicting optimization goals for energy saving and renewable utilization [62]. The fundamental principle of MPC is to predict the future behavior of the controlled system over a finite time horizon and compute an optimal control input that, while ensuring satisfaction of given system constraints, minimizes the priori-defined cost function (as shown in Fig. 2.2). By predicting the weather and electricity prices, Gholamibozanjani et al. [63] developed an MPC to smartly control the space heating system. The simulation results indicate a

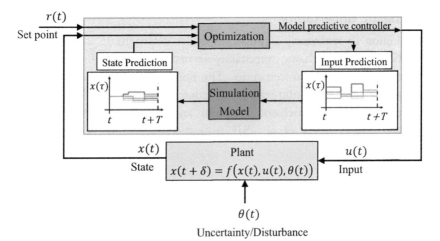

Figure 2.2 Model predictive controller in a feedback loop.

cost saving of about 12%−57%. Sultana et al. [64] comprehensively reviewed MPC on renewable devices, for example, induction motors and various converters. The review can promote the design, analysis, and application of MPC in renewable energy and conversion systems. Zhang et al. [65] applied probabilistic forecasting and rolling stochastic optimization to predict the long-term energy supply of hybrid renewable systems, considering the uncertainty of future climate conditions. Results indicate that the hydropower plant can effectively manage the mismatch between demand and energy supply, while energy storage is required due to fluctuations in renewable energy. Yao and Shekhar [66] provided a review of MPC in HVAC, with respect to prediction and control horizon, time step, and cost function. They concluded that the MPC outperforms others in energy saving, cost saving, and thermal comfort.

In terms of microgrids, Hu et al. [67] provide an overview of MPC in voltage regulation, frequency control, power flow management, and optimization. They concluded that MPC shows superiority in fast responses and flexible system integration. In respect to island microgrids, Zhang et al. [68] developed a two-stage MPC to minimize operation costs and equally allocate economic benefits. Hu et al. [69] applied the MPC strategy to manage the power supply-demand of a PV-Battery microgrid. The state of charge-oriented battery charging enables the reactive capability of microgrids for grid support.

2.4 Future prospects for machine learning in energy transition and sustainable development goals

With the ambitious targets for carbon peaking in 2030 and carbon neutrality in 2060 in China [70,71], ML plays a significant role in energy transition and SDGs. ML is an integrated multienergy system for predictions of both power supply and energy demand to optimize the power dispatch strategy [72]. Furthermore, ML can also be effective for accurate predictions of renewable energy, demand, and dynamic state of battery [73]. Future prospects for ML in the digital energy era include but are not limited to the following:
1. Data mining through knowledge exploitation for performance forecasting.
2. New knowledge exploration through transfer learning is needed to expand the database boundary and improve the robustness of ML.
3. MPCs have trade-off between time consumption and real-time decision-making.

4. Forecasting future national energy structures under carbon neutrality targets, together with pathways on SDG.
5. ML is applied to accelerate advanced material discovery and synthesis for energy conversion and storage [74,75].
6. Automatic fault detection and diagnosis on integrated systems [76].

References

[1] Lee JH, Shin J, Realff MJ. Machine learning: overview of the recent progresses and implications for the process systems engineering field. Computers & Chemical Engineering 2018;114:111−21.
[2] Liu Z, Zhang X, Sun Y, Zhou Y. Advanced controls on energy reliability, flexibility and occupant-centric control for smart and energy-efficient buildings. Energy and Buildings 2023;113436.
[3] Song G. A systematic literature review on smart and personalized ventilation using CO_2 concentration monitoring and control. Energy Reports 2023;8:7523−36.
[4] Liu Z, Sun Y, Xing C, Liu J, He Y, Zhou Y, et al. Artificial intelligence powered large-scale renewable integrations in multi-energy systems for carbon neutrality transition: Challenges and future perspectives. Energy and AI 2022;10:100195.
[5] Zhou Y, Zheng S. Stochastic uncertainty-based optimisation on an aerogel glazing building in China using supervised learning surrogate model and a heuristic optimisation algorithm. Renewable Energy 2020;155:810−26.
[6] Zhou Y, Zheng S. Climate adaptive optimal design of an aerogel glazing system with the integration of a heuristic teaching-learning-based algorithm in machine learning-based optimization. Renewable Energy 2020;153:375−91.
[7] Zhou Y, Zheng S. Uncertainty study on thermal and energy performances of a deterministic parameters based optimal aerogel glazing system using machine-learning method. Energy 2020;193:116718.
[8] Kofinas P, Doltsinis S, Dounis AI, Vouros GA. A reinforcement learning approach for MPPT control method of photovoltaic sources. Renewable Energy 2017;108:461−73.
[9] Shen H, Ruiz A, Li N. Fast online reinforcement learning control of small lift-driven vertical axis wind turbines with an active programmable four bar linkage mechanism. Energy 2023;262:125350.
[10] Zhang Y, Yang X, Liu S. Data-driven predictive control for floating offshore wind turbines based on deep learning and multi-objective optimization. Ocean Engineering 2022;266:112820.
[11] Zhou Y, Liu Z, Zheng S. Influence of novel PCM-based strategies on building cooling performance. Eco-efficient Materials for Reducing Cooling Needs in Buildings and Construction. Woodhead Publishing Series in Civil and Structural Engineering; 2021, p. 329−53.
[12] Perera ATD, Kamalaruban P. Applications of reinforcement learning in energy systems. Renewable and Sustainable Energy Reviews 2021;137:110618.
[13] Ma T, Guo Z, Lin M, Wang Q. Recent trends on nanofluid heat transfer machine learning research applied to renewable energy. Renewable and Sustainable Energy Reviews 2021;138:110494.
[14] Konstantakopoulos IC, Barkan AR, He S, Veeravalli T, Liu H, Spanos C. A deep learning and gamification approach to improving human-building interaction and energy efficiency in smart infrastructure. Applied Energy 2019;237:810−21.

[15] Magazzino C, Mele M, Schneider N. A machine learning approach on the relationship among solar and wind energy production, coal consumption, GDP, and CO_2 emissions. Renewable Energy 2021;167:99−115.

[16] Nam KJ, Hwangbo S, Yoo CK. A deep learning-based forecasting model for renewable energy scenarios to guide sustainable energy policy: a case study of Korea. Renewable and Sustainable Energy Reviews 2020;122:109725.

[17] Leerbeck K, Bacher P, Junker RG, Goranović G, Corradi O, Ebrahimy R, et al. Short-term forecasting of CO_2 emission intensity in power grids by machine learning. Applied Energy 2020;277:115527.

[18] Sharma P, Said Z, Kumar A, et al. Recent advances in machine learning research for nanofluid-based heat transfer in renewable energy system. Energy & Fuels: An American Chemical Society Journal 2022;36(13):6626−58.

[19] Specht JM, Madlener R. Deep reinforcement learning for the optimized operation of large amounts of distributed renewable energy assets. Energy and AI 2023;11:100215.

[20] Izanloo M, Aslani A, Zahedi R. Development of a Machine learning assessment method for renewable energy investment decision making. Applied Energy 2022;327:120096.

[21] Zhou Y, Zheng S, Liu Z, Wen T, Ding Z, Yan J, et al. Passive and active phase change materials integrated building energy systems with advanced machine-learning based climate-adaptive designs, intelligent operations, uncertainty-based analysis and optimisations: A state-of-the-art review. Renewable and Sustainable Energy Reviews 2020;130:109889.

[22] Zhou Y, Liu Z. A cross-scale 'material-component-system' framework for transition towards zero-carbon buildings and districts with low, medium and high-temperature phase change materials. Sustainable Cities and Society 2023;89:104378.

[23] Ahmad T, Chen H. Deep learning for multi-scale smart energy forecasting. Energy 2019;175:98−112.

[24] Xie M, Qiu Y, Liang Y, Zhou Y, Liu Z, Zhang G. Policies, applications, barriers and future trends of building information modeling technology for building sustainability and informatization in China. Energy Reports 2022;8:7107−26.

[25] Zhou Y. Artificial intelligence in renewable systems for transformation towards intelligent buildings. Energy and AI 2022;10:100182.

[26] May Tzuc O, Chan-González JJ, et al. Multivariate inverse artificial neural network to analyze and improve the mass transfer of ammonia in a Plate Heat Exchanger-Type Absorber with NH_3/H_2O for solar cooling applications. Energy Exploration & Exploitation 2022;40(6):1686−711.

[27] Chen SZ, Feng DC, Wang WJ, et al. Probabilistic machine-learning methods for performance prediction of structure and infrastructures through natural gradient boosting. Journal of Structural Engineering 2022;148(8).

[28] Wang H, Lei Z, Zhang X, Zhou B, Peng J. A review of deep learning for renewable energy forecasting. Energy Conversion and Management 2019;198:111799.

[29] Sharifzadeh M, Sikinioti-Lock A, Shah N. Machine-learning methods for integrated renewable power generation: a comparative study of artificial neural networks, support vector regression, and Gaussian Process Regression. Renewable and Sustainable Energy Reviews 2019;108:513−38.

[30] Aslam S, Herodotou H, Mohsin SM, Javaid N, Ashraf N, Aslam S. A survey on deep learning methods for power load and renewable energy forecasting in smart microgrids. Renewable and Sustainable Energy Reviews 2021;144:110992.

[31] Liu H, Chen C, Lv X, Wu X, Liu M. Deterministic wind energy forecasting: a review of intelligent predictors and auxiliary methods. Energy Conversion and Management 2019;195:328−45.

[32] Lin Z, Liu X. Wind power forecasting of an offshore wind turbine based on high-frequency SCADA data and deep learning neural network. Energy 2020;201:117693.

[33] Hong YY, Rioflorido CLPP. A hybrid deep learning-based neural network for 24-h ahead wind power forecasting. Applied Energy 2019;250:530−9.

[34] Mellit A, Pavan AM, Lughi V. Deep learning neural networks for short-term photo-voltaic power forecasting. Renewable Energy 2021;172:276−88.

[35] Voyant C, Notton G, Kalogirou S, Nivet ML, Paoli C, Motte F, et al. Machine learning methods for solar radiation forecasting: a review. Renewable Energy 2017;105:569−82.

[36] Wang K, Qi X, Liu H. A comparison of day-ahead photovoltaic power forecasting models based on deep learning neural network. Applied Energy 2019;251:113315.

[37] Qin W, Wang L, Lin A, Zhang M, Xia X, Hu B, et al. Comparison of deterministic and data-driven models for solar radiation estimation in China. Renewable and Sustainable Energy Reviews 2018;81:579−94.

[38] Lee J, Hong S, Cho H, Lyu B, Kim M, Kim J, et al. Machine learning-based energy optimization for on-site SMR hydrogen production. Energy Conversion and Management 2021;244:114438.

[39] Hwangbo S, Nam KJ, Heo SK, Yoo CK. Hydrogen-based self-sustaining integrated renewable electricity network (HySIREN) using a supply-demand forecasting model and deep-learning algorithms. Energy Conversion and Management 2019;185:353−67.

[40] Long B, Fischer B, Zeng Y, et al. Machine learning-informed and synthetic biology-enabled semi-continuous algal cultivation to unleash renewable fuel productivity. Nature Communications 2022;13:541.

[41] Potnuri R, Suriapparao DV, Rao CS, Sridevi V, Kumar A. Effect of dry torrefaction pretreatment of the microwave-assisted catalytic pyrolysis of biomass using the machine learning approach. Renewable Energy 2022;197:798−809.

[42] Naqvi SR, Ullah Z, Taqvi SAA, Khan MNA, Farooq W, et al. Applications of machine learning in thermochemical conversion of biomass—a review. Fuel 2023;332:126055.

[43] Zhou Y, Zheng S, Zhang G. Machine-learning based study on the on-site renewable electrical performance of an optimal hybrid PCMs integrated renewable system with high-level parameters' uncertainties. Renewable Energy 2020;151:403−18.

[44] Wang Y, Hu Q, Meng D, Zhu P. Deterministic and probabilistic wind power fore-casting using a variational Bayesian-based adaptive robust multi-kernel regression model. Applied Energy 2017;208:1097−112.

[45] Mashlakov A, Kuronen T, Lensu L, Kaarna A, Honkapuro S. Assessing the performance of deep learning models for multivariate probabilistic energy forecasting. Applied Energy 2021;285:116405.

[46] Wang H, Yi H, Peng J, Wang G, Liu Y, Jiang H, et al. Deterministic and probabilistic forecasting of photovoltaic power based on deep convolutional neural network. Energy Conversion and Management 2017;153:409−22.

[47] Zhang Y, Liu K, Qin L, An X. Deterministic and probabilistic interval prediction for short-term wind power generation based on variational mode decomposition and machine learning methods. Energy Conversion and Management 2016;112:208−19.

[48] Xie Y, Li C, Li M, Liu F, Taukenova M. An overview of deterministic and probabilistic forecasting methods of wind energy. iScience 2022;105804.

[49] Shen R, Zhong S, Wen X, An Q, Zheng R, Li Y, et al. Multi-agent deep reinforcement learning optimization framework for building energy system with renewable energy. Applied Energy 2022;312:118724.

[50] Harrold DJB, Cao J, Fan Z. Renewable energy integration and microgrid energy trading using multi-agent deep reinforcement learning. Applied Energy 2022;318:119151.

[51] Zhang B, Hu W, Cao D, Huang Q, Chen Z, Blaabjerg F. Deep reinforcement learning−based approach for optimizing energy conversion in integrated electrical and heating system with renewable energy. Energy Conversion and Management 2019;202:112199.

[52] Zhou Y, Zheng S, Zhang G. Machine learning-based optimal design of a phase change material integrated renewable system with on-site PV, radiative cooling and hybrid ventilations—study of modelling and application in five climatic regions. Energy 2020;192:116608.

[53] Wang W, Ma Y, Maroufmashat A, Zhang N, Li J, Xiao X. Optimal design of large-scale solar-aided hydrogen production process via machine learning based optimisation framework. Applied Energy 2022;305:117751.

[54] Gao Y, Matsunami Y, Miyata S, Akashi Y. Operational optimization for off-grid renewable building energy system using deep reinforcement learning. Applied Energy 2022;325:119783.

[55] Yi Z, Luo Y, Westover T, Katikaneni S, Ponkiya B, et al. Deep reinforcement learning based optimization for a tightly coupled nuclear renewable integrated energy system. Applied Energy 2022;328:120113.

[56] Sahinidis NV. Optimization under uncertainty: state-of-the-art and opportunities. Computers & Chemical Engineering 2004;28(6−7):971−83.

[57] Zakaria A, Ismail FB, Lipu MSH, Hannan MA. Uncertainty models for stochastic optimization in renewable energy applications. Renewable Energy 2020;145:1543−71.

[58] Guevara E, Babonneau F, Homem-de-Mello T, Moret S. A machine learning and distributionally robust optimization framework for strategic energy planning under uncertainty. Applied Energy 2020;271:115005.

[59] Crespo-Vazquez JL, Carrillo C, Diaz-Dorado E, Martinez-Lorenzo JA, Noor-E-Alam M. A machine learning based stochastic optimization framework for a wind and storage power plant participating in energy pool market. Applied Energy 2018;232:341−57.

[60] Zhou Y, Zheng S. Multi-level uncertainty optimisation on phase change materials integrated renewable systems with hybrid ventilations and active cooling. Energy 2020;202:117747.

[61] Zhou Y. A multi-stage supervised learning optimisation approach on an aerogel glazing system with stochastic uncertainty. Energy 2022;258:124815.

[62] Killian M, Kozek M. Ten questions concerning model predictive control for energy efficient buildings. Building and Environment 2016;105:403−12.

[63] Gholamibozanjani G, Tarragona J, De Gracia A, Fernández C, Cabeza LF, Farid MM. Model predictive control strategy applied to different types of building for space heating. Applied Energy 2018;231:959−71.

[64] Sultana WR, Sahoo SK, Sukchai S, Yamuna S, Venkatesh D. A review on state of art development of model predictive control for renewable energy applications. Renewable and Sustainable Energy Reviews 2017;76:391−406.

[65] Zhang Y, Cheng C, Cai H, Jin X, Jia Z, Wu X, et al. Long-term stochastic model predictive control and efficiency assessment for hydro-wind-solar renewable energy supply system. Applied Energy 2022;316:119134.

[66] Yao Y, Shekhar DK. State of the art review on model predictive control (MPC) in heating ventilation and air-conditioning (HVAC) field. Building and Environment 2021;200:107952.

[67] Hu J, Shan Y, Guerrero JM, Ioinovici A, Chan KW, Rodriguez J. Model predictive control of microgrids − an overview. Renewable and Sustainable Energy Reviews 2021;136:110422.

[68] Zhang Y, Fu L, Zhu W, Bao X, Liu C. Robust model predictive control for optimal energy management of island microgrids with uncertainties. Energy 2018;164:1229−41.

[69] Hu J, Xu Y, Cheng KW, Guerrero JM. A model predictive control strategy of PV-Battery microgrid under variable power generations and load conditions. Applied Energy 2018;221:195−203.

[70] Liu Z, Yu C, Qian K, Huang R, You K, Henk V, et al. Incentive initiatives on energy-efficient renovation of existing buildings towards carbon−neutral blueprints in China: Advancements, challenges and prospects. Energy and Buildings 2023;296:113343.

[71] Liu Z, Zhou Y, Yan J, Marcos T. Frontier ocean thermal/power and solar PV systems for transformation towards net-zero communities. Energy 2023;284:128362.

[72] Shivam K, Tzou JC, Wu SC. A multi-objective predictive energy management strategy for residential grid-connected PV-battery hybrid systems based on machine learning technique. Energy Conversion and Management 2021;237:114103.

[73] Dan Z, Song A, Yu X, Zhou Y. An artificial intelligence-based modelling approach in a circular economy with E-mobility-based interactive renewable sharing. Renewable & Sustainable Energy Reviews 2023; Under review.

[74] ChenA., ZhangX., ZhouZ.. Machine learning: accelerating materials development for energy storage and conversion. InfoMat 2020;2(3):553−76.

[75] Liu Y, Esan OC, Pan Z, An L. Machine learning for advanced energy materials. Energy and AI 2021;3:100049.

[76] Manno D, Cipriani G, Ciulla G, Dio VD, Guarino S, Lo Brano V. Deep learning strategies for automatic fault diagnosis in photovoltaic systems by thermographic images. Energy Conversion and Management 2021;241:114315.

CHAPTER 3

Building occupant behavior and vehicle driving schedules with demand prediction and analysis

Yuekuan Zhou[1,2,3,4]
[1]Sustainable Energy and Environment Thrust, Function Hub, The Hong Kong University of Science and Technology (Guangzhou), Nansha, Guangdong, P.R. China
[2]Department of Mechanical and Aerospace Engineering, The Hong Kong University of Science and Technology, Clear Water Bay, Hong Kong SAR, P.R. China
[3]HKUST Shenzhen-Hong Kong Collaborative Innovation Research Institute, Futian, Shenzhen, P.R. China
[4]Division of Emerging Interdisciplinary Areas, The Hong Kong University of Science and Technology, Clear Water Bay, Hong Kong SAR, P.R. China

3.1 Introduction

The ever-increasing contradiction between global warming from fossil fuel consumption and the energy shortage crisis calls for a significant renewable energy supply with efficient utilization [1−3]. Due to the decrease in power transmission losses, onsite self-consumption of distributed renewable energy (DRE) plays a significant role in the carbon neutrality transition through large-scale renewable integrations [4], demand-side control [5], future climate prediction, and energy system planning [6]. However, restrained by intermittence and fluctuation, the spatiotemporal mismatch between energy supply and demand imposes challenges on direct self-consumption [7−9]. Advanced modeling techniques for building occupant behavior and vehicle driving schedules are necessary for accurate demand prediction and analysis, enabling day-ahead power dispatch on renewable energy and charging/discharging of storage.

In academia, researchers have focused on accurate demand prediction and energy analytics by developing mathematical models of occupant behavior and electric vehicle (EV) driving schedules. In terms of indoor occupant behavior, various models are involved, for example, Stochastic Markov models [10], data mining approaches [11], data-driven models [12], and so on. With respect to the arrival, detention, and departure schedules of vehicles, the mathematical models mainly include normal distribution, Chi-square distribution [13], etc. Yi et al. [14] developed a spatial-temporal distribution model with Monte Carlo simulation for charging demand prediction of EVs. Based on the predicted demand, a clear charging pile distribution strategy is worked out as a slow-to-fast charging pile ratio of 50:1 in residential areas, 85:1 in working areas, and 3:1 in other functional areas. In terms of EV driving scheduling, He et al. [15] conducted a bi-layer optimization of spatiotemporal scheduling on EVs with respect to the time domain and spatial location of the EV load. The EV charging/discharging load has a great impact on distribution network planning. To timely provide charging services to EVs, Xu and Huang [16] developed a hybrid clustering algorithm and a charging pile allocation model. The testing results indicate that a 25% cost reduction can be achieved.

Following indoor occupant behavior and vehicle driving schedules, the spatiotemporal distribution of charging pile allocation and smart charging/discharging with renewable dispatch has also attracted widespread interest. Pareschi et al. [17] studied factors for EV charging, based on empirical EV travel survey data. They indicated that critical factors include charging power, efficiency, battery size, and battery state of charge. With respect to the uncertain profitability of charging infrastructures given the low penetration of EVs, Pagani et al. [18] developed an integrated simulation platform to optimize the allocation of EV charging infrastructure. Dixon and Bell [19] studied battery capacity and charging access strategies to mitigate grid power pressure and shift peak domestic demand. Yap et al. [20] reviewed cutting-edge technologies in solar-powered battery electric vehicle charging stations. They concluded that the main technical issues include solar energy intermittency, slow charging speed, blackout of the power grid, and uncontrollability of grid charging in the periods with high carbon emission factors. The solutions to addressing these issues include hybrid wind or biogas integrations, battery swapping services, virtual inertia devices for frequency regulation, and decentralized energy management [21].

As an auxiliary energy storage device, mobile EVs show great potential in multidirectional energy interactions. Zhou [22] studied techno-economic analysis on an interactive energy network within the Greater Bay Area. The comparison between vehicle-to-building (V2B) and vehicle-to-grid (V2G) indicated that V2G overwhelms V2B by increasing the internal trading cost to 1.78×10^5 HK\$, which is much higher than the battery depreciation cost of 0.688×10^5 HK\$. To actively incentivize multistakeholders' proactivity and market vitality,

Zhou [23] developed an integrated platform for system design, planning, and operation, together with policy-making and energy management. The developed platform can incentivize the mutual win-to-win collaboration for sustainability transformation. Robledo et al. [24] studied combined heat and power supply in PV (photovoltaic)-residential building fuel cell EVs (FCEVs). Results indicate that the FCEV with V2G mode can reduce grid import power by around 71%. Barone et al. [25] studied a building-vehicle-building paradigm for renewable penetration and grid independence.

3.2 Comprehensive literature review on occupant behavior, vehicle driving schedules, and energy analytics

3.2.1 Mathematical modeling on stochastic occupant behavior and energy analytics in buildings

To characterize the stochastic behaviors of human beings in building energy use and EV driving, researchers have developed mathematical models based on both theoretical and onsite questionnaire databases. Table 3.1 summarizes modeling on occupant behavior. Hong et al. [29] comprehensively reviewed the uncertainty of occupant behavior in building energy prediction and pointed out three main challenges involving data collection techniques, analytical and modeling methods, and simulation applications. To inspire innovative research and applications, Hong et al. [30] summarized ten questions and clarified the technical challenges, from the perspectives of concepts, applications, and methodologies. Dong et al. [31] comprehensively reviewed occupant behavior modeling methods in urban scales, including stochastic modeling, neural networks, reinforcement learning, and network modeling. Various metrics can be applied for model accuracy evaluation in different application scenarios. The methodological framework in IEA EBC Annex 66 [32] includes data collection, modeling and evaluation approaches, and modeling integration tools. Jia et al. [33] comprehensively reviewed modeling approaches on occupant behavior, together with an in-depth analysis of the advantages and limitations of each model.

3.2.2 Vehicle driving schedules and energy interactions in renewable-grid-load-storage framework

Fig. 3.1 shows an integrated renewable-grid-load-storage framework. Cleaner power from solar PVs and wind turbines fulfills the demands of

Table 3.1 A holistic overview on modeling of occupant behavior and energy analytics in buildings.

Studies	Parameters	Methods	Results
Haldi and Robinson [26]	Deterministic and stochastic occupant presence and behavior	Urban energy modeling tool, urban energy modeling tool	Occupant behavior and individual diversity show significant impact on building energy demand
Gaetani et al. [27]	Building construction parameters (window type, wall construction, and roof construction), operation parameters (heating, ventilation and air-conditioning (HVAC) schedules, setpoint for cooling/ heating, occupancy schedule and rate, lighting schedule, and power density)	A fit-for-purpose model	Most suitable model for occupant behavior is case-dependent
Virote and Neves-Silva [10]	Human behavior	Stochastic Markov models	The model can accurately learn occupant behavioral patterns, predict energy consumption, and identify energy-saving areas.

(*Continued*)

Table 3.1 (Continued)

Studies	Parameters	Methods	Results
Yu et al. [28]	City climate, building-related characteristics, user-related characteristics, and building services systems	Data mining for occupant behavior analysis and min−max normalization for data consistency	Results can optimize occupant behavior for building energy saving.
Zhao et al. [11]	Occupancy schedule and occupancy rate	Data mining approach for occupant behavior learning	Impact of occupancy schedules on HVAC energy consumption is climate-dependent.
Piselli and Pisello [12]	Standard occupancy and field-collected data	Data-driven models	20% discrepancy can be noticed between simulation and collected results.

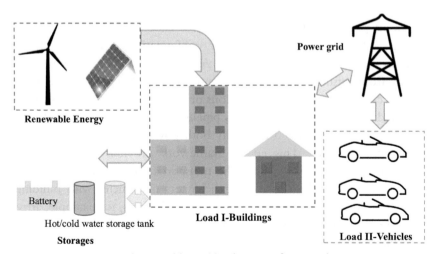

Figure 3.1 An integrated renewable-grid-load-storage framework.

buildings and vehicles. The mismatch will charge storages. The power grid works as a backup to dynamically balance supply and demand. As one of the most critical parameters, vehicle schedules (i.e., parking, detention, and departure) will significantly affect real-time renewable penetration and demand coverage. Zhou [34] studied the impact of different stochastic vehicle schedule models on techno-economic-environmental performance. Results showed that the normal distribution of arriving time and detention time-duration of vehicles show the minimum annual import, while the normal distribution of arriving time and departure time with a Chi-square distribution for detention time-duration show the minimum equivalent CO_2 emission. Thomas et al. [13] assumed that the arrival time of EVs follows the chi-square distribution, and the detention time of the EVs follows the normal distribution.

Table 3.2 comprehensively reviews vehicle driving schedules and energy interactions. Considering the significant impact of EV driving schedules on the energy sharing of DRE systems, Cardoso et al. [38] studied the optimal DRE under uncertainties of EV driving patterns. Results indicate that for short payback periods, EV driving schedules have a high impact on optimal DER decisions. Karan et al. [39] adopted stochastic optimization to decarbonize the CO_2 emissions of integrated buildings and transportation. Results showed a 4% and 11% reduction in CO_2 emissions for grid-tied and off-grid systems, respectively. Alinejad et al. [40] explored optimal charging/discharging schedules under the stochastic behavior of EVs in parking lots, that is, arrival, departure, and detention time. The optimal schedule refers to shifting EV charging from peak periods to off-peak periods.

Fig. 3.2 demonstrates a systematic configuration in a coastal district with an expanded energy boundary. As shown in Fig. 3.2A, renewable energy from offshore wind turbines and floating PV farms can be either used to charge EVs in charging stations or converted to H_2, which can be delivered to H_2 stations through H_2 pipelines. Both EVs and HVs (hybrid vehicles) can be discharged to supply both power and thermal energy to district communities. This energy framework provides a clean energy transition and decarbonizes both the transportation and building sectors. Note that the charging/discharging behaviors of vehicles with buildings and stations are highly dependent on vehicle schedules, for example, arriving, detention, and departure time. In academia, different probability density functions have been studied, while the main difficulty is imposed by the stochastic characteristics of driving behaviors.

Table 3.2 Review of vehicle driving schedules and energy interactions.

Studies	Systems	Methods	Results
Mehrjerdi and Hemmati [35]	Wind energy-electric vehicle charging station-energy storage	Stochastic mixed integer linear programming	15% and 12% of the total cost can be covered by network reinforcement and charging facilities.
Thomas et al. [13]	Building EV grid	Stochastic electric vehicle driving schedules	Daily cost is much lower in stochastic cases than that in deterministic cases.
Fischer et al. [36]	Building EVs	Inhomogeneous Markov-chain stochastic modeling approach	Uncontrolled EV charging will increase load peaks by up to 8.5 times together with the annual electricity demand by two times.
Xu et al. [37]	PV–EV-a commercial building	Stochastic behaviors based on a mixed-integer linear programming model.	Operational costs can be reduced by over 80%.

EV, Electric vehicle; PV, Photovoltaic.

3.3 Energy interaction mechanisms in integrated multienergy systems with economic feasibility

With respect to an integrated multienergy system (MES) with EVs and battery energy storage systems (BESSs), Zeynali et al. [41] developed a two-stage stochastic EMS, considering different charging rates of EV and BESS. MESs involve interactions in electricity, heat, cooling, and fuels at

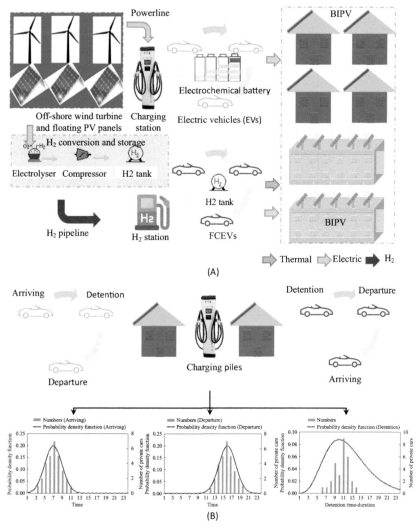

Figure 3.2 Systematic configuration with expanded energy boundary: (A) integrated multi-energy system in energy districts; (B) driving schedules of vehicles with probability density functions.

different levels [42]. Klemm and Vennemann [43] comprehensively reviewed modeling techniques for MES in mixed-use districts and concluded that there are few models available. In an integrated MES, thermal/electrical/hydrogen energy can be shared between building services systems and new energy vehicles, especially with electrification and hydrogenation transformations [44]. Multidirectional energy interactions can be achieved, including B2V (building-to-vehicle), V2B, V2G, V2G (grid-to-vehicle), etc. [45]. By transforming from internal combustion engine vehicles to battery EVs and from gas boilers to heat pumps, Murray et al. [15] conducted multiobjective optimization on a decentralized MES with electrification of buildings and vehicles. Results indicated that in the tested 77 vehicles and 50 buildings, CO_2 emissions could be reduced by 79% in 2035 and by 85% in 2050.

From the lifecycle perspective, Zhou and Cao [46] investigated the techno-economic performance of a battery cycling, aging-based energy management strategy. Results indicated that in the EV-based energy-sharing framework, the economic subsidy on the grid feed-in tariff is quite necessary for economic feasibility. Furthermore, the transition from the negative to the positive energy district with renewable energy sharing and flexible energy management shows an increase in net present value [47].

In addition to EVs, hydrogen-based thermal/power energy sharing has also attracted researchers' interest. He et al. [48] studied hydrogen economy sharing between residential communities and hydrogen vehicles. For grid stability and annual cost saving, the proposed energy dispatch strategy can reduce the maximum mean hourly grid power to 78.2 kW by 24.2% and the annual energy cost to 1228.5 $/household by 38.9%. The quantification of fuel cell degradation [49] requires high economic incentives to support the hydrogen-based sharing framework. Mehrjerdi et al. [50] studied power control and energy management strategies for grid-connected buildings with hydrogen vehicles and wind-solar-battery. The operating mechanism is to charge EVs and HVs with renewable energy and discharge storage for demand coverage.

Fig. 3.3 shows an overview framework of a cross-regional energy community, consisting of EV charging piles, distributed solar PVs, and onsite wind turbines, together with P2P energy sharing, B2V/V2B, and G2B energy interactions. In addition to power transmission lines, the mobility of EVs can manage the spatiotemporal energy balance to maximize renewable use. Furthermore, instead of being exported to the local power grid, renewable energy can be self-consumed, stabilizing the local power grid.

3.4 Advanced modeling techniques and guidelines for techno-economic-environmental performance improvement

With respect to building occupant behavior and vehicle driving schedules with demand prediction and analysis, advanced modeling techniques and

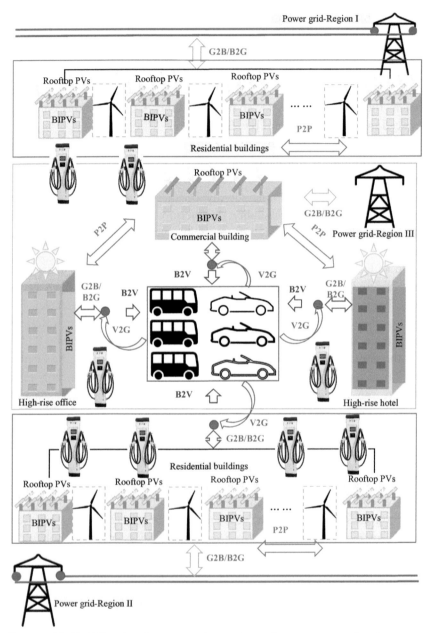

Figure   3.3 Multidirectional   energy   interactions   in   a   cross-regional   energy community.

guidelines are necessary for integrated system design and operation. Future studies can be focused on:

1. Subsidy in grid feed-in tariff to support the sustainability transformation, including distributed renewable systems, E-transportation, and battery-based power sharing.
2. Wireless power transfer with high efficiency.
3. Battery-based circular economy and waste-to-energy economy.
4. Energy equality and cost allocation.
5. Consciousness enhancement in residents to actively participate in the energy sharing framework with synergistic building occupant behavior and vehicle driving schedules.
6. Spatiotemporal allocation and optimal distribution of charging piles to improve renewable penetration and address cruise anxiety simultaneously.

References

[1] Liu Z, Xie M, Zhou Y, He Y, Zhang L, Zhang G, et al. A state-of-the-art review on shallow geothermal ventilation systems with thermal performance enhancementsingle bond system classifications, advanced technologies and applications. Energy and Built Environment 2023;4(2):148−68.
[2] Liu Z, Yu C, Qian K, Huang R, You K, Henk V, et al. Incentive initiatives on energy-efficient renovation of existing buildings towards carbon−neutral blueprints in China: Advancements, challenges and prospects. Energy and Buildings 2023;296:113343.
[3] Zhou Y, Liu Z. A cross-scale 'material-component-system' framework for transition towards zero-carbon buildings and districts with low, medium and high-temperature phase change materials. Sustainable Cities and Society 2023;89:104378.
[4] Liu Z, Sun Y, Xing C, Liu J, He Y, Zhou Y, et al. Artificial intelligence powered large-scale renewable integrations in multi-energy systems for carbon neutrality transition: challenges and future perspectives. Energy and AI 2022;10:100195.
[5] Zhou Y, Zheng S. Machine-learning based hybrid demand-side controller for high-rise office buildings with high energy flexibilities. Applied Energy 2020;262:114416.
[6] Zhou Y. Energy planning and advanced management strategies for an interactive zero-energy sharing network (buildings and electric vehicles) with high energy flexibility and electrochemical battery cycling aging. Hong Kong Polytechnic University [Dissertations], 2021;.
[7] Liu Z, Zhang X, Sun Y, Zhou Y. Advanced controls on energy reliability, flexibility, resilience, and occupant-centric control for smart and energy-efficient buildings. Energy and Buildings 2023;297:113436.
[8] Sun Y, Fariborz H, Benjamin C. A review of the-state-of-the-art in data-driven approaches for building energy prediction. Energy and Buildings 2020;221:110022.
[9] Sun Y, Benjamin C, Fariborz H. In-Processing fairness improvement methods for regression Data-Driven building Models: Achieving uniform energy prediction. Energy and Buildings 2022;277:112565.

[10] Virote J, Neves-Silva R. Stochastic models for building energy prediction based on occupant behavior assessment. Energy and Buildings 2012;53:183−93.

[11] Zhao J, Lasternas B, Lam KP, Yun R, Loftness V. Occupant behavior and schedule modeling for building energy simulation through office appliance power consumption data mining. Energy and Buildings 2014;82:341−55.

[12] Piselli C, Pisello AL. Occupant behavior long-term continuous monitoring integrated to prediction models: Impact on office building energy performance. Energy 2019;176:667−81.

[13] Thomas D, Deblecker O, Ioakimidis CS. Optimal operation of an energy management system for a grid-connected smart building considering photovoltaics' uncertainty and stochastic electric vehicles' driving schedule. Applied Energy 2018;210:1188−206.

[14] Yi T, Zhang C, Lin T, Liu J. Research on the spatial-temporal distribution of electric vehicle charging load demand: a case study in China. Journal of Cleaner Production 2020;242:118457.

[15] He L, Yang J, Yan J, Tang Y, He H. A bi-layer optimization based temporal and spatial scheduling for large-scale electric vehicles. Applied Energy 2016;168:179−92.

[16] Xu J, Huang Y. The short-term optimal resource allocation approach for electric vehicles and V2G service stations. Applied Energy 2022;319:119200.

[17] Pareschi G, Küng L, Georges G, Boulouchos K. Are travel surveys a good basis for EV models? Validation of simulated charging profiles against empirical data. Applied Energy 2020;275:115318.

[18] Pagani M, Korosec W, Chokani N, Abhari RS. User behaviour and electric vehicle charging infrastructure: an agent-based model assessment. Applied Energy 2019;254:113680.

[19] Dixon J, Bell K. Electric vehicles: Battery capacity, charger power, access to charging and the impacts on distribution networks. eTransportation 2020;4:100059.

[20] Yap KY, Chin HH, Klemeš JJ. Solar energy-powered battery electric vehicle charging stations: current development and future prospect review. Renewable and Sustainable Energy Reviews 2022;169:112862.

[21] Liu Z, Zhou Y, Yan J, Marcos T. Frontier ocean thermal/power and solar PV systems for transformation towards net-zero communities. Energy 2023;284:128362.

[22] Zhou Y. Energy sharing and trading on a novel spatiotemporal energy network in Guangdong-Hong Kong-Macao Greater Bay Area. Applied Energy 2022;318:119131.

[23] Zhou Y. Incentivising multi-stakeholders' proactivity and market vitality for spatiotemporal microgrids in Guangzhou-Shenzhen-Hong Kong Bay Area. Applied Energy 2022;328:120196.

[24] Robledo CB, Oldenbroek V, Abbruzzese F, Wijk AJM. Integrating a hydrogen fuel cell electric vehicle with vehicle-to-grid technology, photovoltaic power and a residential building. Applied Energy 2018;215:615−29.

[25] Barone G, Buonomano A, Calise F, Forzano C, Palombo A. Building to vehicle to building concept toward a novel zero energy paradigm: modelling and case studies. Renewable and Sustainable Energy Reviews 2019;101:625−48.

[26] Haldi F, Robinson D. The impact of occupants' behaviour on building energy demand. Journal of Building Performance Simulation 2011;4(4):323−38.

[27] Gaetani I, Hoes PJ, Hensen JLM. Occupant behavior in building energy simulation: towards a fit-for-purpose modeling strategy. Energy and Buildings 2016;121:188−204.

[28] Yu Z, Fung BCM, Haghighat F, Yoshino H, Morofsky E. A systematic procedure to study the influence of occupant behavior on building energy consumption. Energy and Buildings 2011;43(6):1409−17.

[29] Hong T, Taylor-Lange SC, D'Oca S, Yan D, Corgnati SP. Advances in research and applications of energy-related occupant behavior in buildings. Energy and Buildings 2016;116:694−702.

[30] Hong T, Yan D, D'Oca S, Chen C. Ten questions concerning occupant behavior in buildings: the big picture. Building and Environment 2017;114:518−30.

[31] Dong B, Liu Y, Fontenot H, et al. Occupant behavior modeling methods for resilient building design, operation and policy at urban scale: a review. Applied Energy 2021;293:116856.

[32] Yan D, Hong T, Dong B, et al. IEA EBC Annex 66: definition and simulation of occupant behavior in buildings. Energy and Buildings 2017;156:258−70.

[33] Jia M, Srinivasan RS, Raheem AA. From occupancy to occupant behavior: an analytical survey of data acquisition technologies, modeling methodologies and simulation coupling mechanisms for building energy efficiency. Renewable and Sustainable Energy Reviews 2017;68:525−40.

[34] Zhou Y. A stochastic vehicle schedule model for a renewable-building-e-transportation-microgrid energy sharing paradigm. Applied Energy 2022; Under review.

[35] Mehrjerdi H, Hemmati R. Stochastic model for electric vehicle charging station integrated with wind energy. Sustainable Energy Technologies and Assessments 2020;37:100577.

[36] Fischer D, Harbrecht A, Surmann A, McKenna R. Electric vehicles' impacts on residential electric local profiles—a stochastic modelling approach considering socio-economic, behavioural and spatial factors. Applied Energy 2019;233−234:644−58.

[37] Xu X, Hu W, Liu W, Du Y, Huang R, Huang Q, et al. Risk management strategy for a renewable power supply system in commercial buildings considering thermal comfort and stochastic electric vehicle behaviors. Energy Conversion and Management 2021;230:113831.

[38] Cardoso G, Stadler M, Bozchalui MC, Sharma R, Marnay C, Barbosa-Póvoa A, et al. Optimal investment and scheduling of distributed energy resources with uncertainty in electric vehicle driving schedules. Energy 2014;64:17−30.

[39] Karan E, Asadi S, Ntaimo L. A stochastic optimization approach to reduce greenhouse gas emissions from buildings and transportation. Energy 2016;106:367−77.

[40] Alinejad M, Rezaei O, Kazemi A, Bagheri S. An optimal management for charging and discharging of electric vehicles in an intelligent parking lot considering vehicle owner's random behaviors. Journal of Energy Storage 2021;35: 102245.

[41] Zeynali S, Rostami N, Ahmadian A, Elkamel A. Two-stage stochastic home energy management strategy considering electric vehicle and battery energy storage system: an ANN-based scenario generation methodology. Sustainable Energy Technologies and Assessments 2020;39:100722.

[42] Mancarella P. MES (multi-energy systems): an overview of concepts and evaluation models. Energy 2014;65:1−17.

[43] Klemm C, Vennemann P. Modeling and optimization of multi-energy systems in mixed-use districts: a review of existing methods and approaches. Renewable and Sustainable Energy Reviews 2021;135:110206.

[44] Zhou Y. Transition towards carbon-neutral districts based on storage techniques and spatiotemporal energy sharing with electrification and hydrogenation. Renewable and Sustainable Energy Reviews 2022;162:112444.

[45] Zhou Y, Cao S, Hensen JLM, Lund PD. Energy integration and interaction between buildings and vehicles: a state-of-the-art review. Renewable and Sustainable Energy Reviews 2019;114:109337.

[46] Zhou Y, Cao S. Coordinated multi-criteria framework for cycling aging-based battery storage management strategies for positive building−vehicle system with renewable depreciation: life-cycle based techno-economic feasibility study. Energy Conversion and Management 2020;226:113473.

[47] Zhou Y, Cao S, Hensen JLM. An energy paradigm transition framework from negative towards positive district energy sharing networks—battery cycling aging, advanced battery management strategies, flexible vehicles-to-buildings interactions, uncertainty and sensitivity analysis. Applied Energy 2021;288:116606.

[48] He Y, Zhou Y, Yuan J, Liu Z, Wang Z, Zhang G. Transformation towards a carbon-neutral residential community with hydrogen economy and advanced energy management strategies. Energy Conversion and Management 2021;249:114834.

[49] He Y, Zhou Y, Wang Z, Liu J, Liu Z, Zhang G. Quantification on fuel cell degradation and techno-economic analysis of a hydrogen-based grid-interactive residential energy sharing network with fuel-cell-powered vehicles. Applied Energy 2021;303:117444.

[50] Mehrjerdi H, Bornapour M, Hemmati R, Ghiasi SMS. Unified energy management and load control in building equipped with wind-solar-battery incorporating electric and hydrogen vehicles under both connected to the grid and islanding modes. Energy 2019;168:919—30.

CHAPTER 4

Integrated energy flexible building and e-mobility with demand-side management and model predictive control

Zhaohui Dan[1] and Yuekuan Zhou[1,2,3,4]
[1]Sustainable Energy and Environment Thrust, Function Hub, The Hong Kong University of Science and Technology (Guangzhou), Nansha, Guangdong, P.R. China
[2]Department of Mechanical and Aerospace Engineering, The Hong Kong University of Science and Technology, Clear Water Bay, Hong Kong SAR, P.R. China
[3]HKUST Shenzhen-Hong Kong Collaborative Innovation Research Institute, Futian, Shenzhen, P.R. China
[4]Division of Emerging Interdisciplinary Areas, The Hong Kong University of Science and Technology, Clear Water Bay, Hong Kong SAR, P.R. China

4.1 Introduction to integrated energy flexible building and e-mobility system

Buildings account for a large share of global energy consumption, accounting for about 30%−45% of the total energy consumption [1,2]. The huge building energy usage is mainly due to various energy-consuming subsystems, including lighting systems; heating, ventilation, and air conditioning (HVAC) systems; water heaters; and other electrical equipment driven to meet the functional, health, and comfort requirements of occupants [3−5]. To promote building energy efficiency and carbon reduction, renewable energy sources, such as wind turbines and photovoltaic (PV) modules, including, rooftop PVs and building-integrated photovoltaics (BIPV), are being studied and applied to provide clean energy to buildings on-site [6,7]. However, with the increasing dynamics of building power demand and the proportion of intermittent renewable energy generation, it is challenging for the grid to maintain a real-time energy supply and demand balance [8]. Energy flexible buildings with controllable and adjustable electrical devices are being studied and implemented in demand-side management (DSM) to enhance energy management [9,10]. Electric vehicles (EVs) can further improve building energy flexibility through their mobility and electricity storage capacity [11−13].

Advances in Digitalization and Machine Learning for Integrated
Building-Transportation Energy Systems
DOI: https://doi.org/10.1016/B978-0-443-13177-6.00011-4

The configuration of a typical integrated energy flexible building and e-mobility system, and building energy flexibility sources, and energy interactions in this system are introduced in detail in the following sections.

4.1.1 Configuration of an integrated building and e-mobility system

Fig. 4.1 shows the architecture of an integrated building and e-mobility system, where there are three types of buildings, including a residential building, an office building, and a commercial building, and a variety of EVs for transportation and freight, such as electric shuttle buses and private EVs. Each building in the system is assumed to be powered by a renewable energy generation system, including BIPVs, on-site wind turbines and static storage

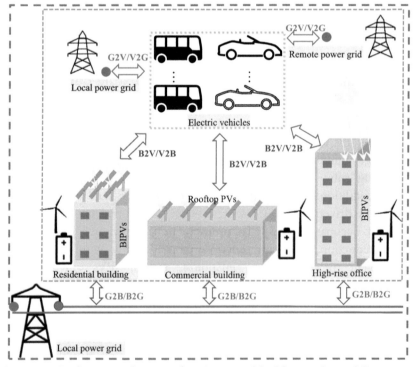

Figure 4.1 Configuration diagram of an integrated building and e-mobility system with interactions. (Note: G2V/V2G refers to the electric vehicle charging from the grid and discharging to the grid, respectively; B2V/V2B refers to the electric vehicle charging from the building power generation system and discharging to the building batteries, respectively; G2B/B2G refers to the building energy system importing electricity from the grid and exporting to the grid, respectively.)

batteries, and the local power grid. As shown in Fig. 4.1, there are multiple energy interaction modes among buildings, EVs, and grids, which provide the possibility of efficient utilization of renewable energy and cross-regional energy sharing [14–17]. Furthermore, with technologies of smart devices and smart meters, energy-efficient buildings are possible. The measurable and controllable building EV energy interaction process increases building energy flexibility and provides a potential method for building energy DSM [18].

The intermittency and uncertainty of wind energy and solar energy make it difficult to achieve real-time energy balance and restrain renewable energy utilization [19]. Providing storage capacity to buffer the fluctuation of wind and solar generation is effective, and it can help reduce energy costs in response to time-of-use tariffs [20]. Battery energy storage is a high-energy storage device that can act as a buffer to store surplus power generation and provide electricity to the building when needed [20]. Electricity storage is used most of the time as a demand regulator or energy buffer during the day to achieve peak shaving, frequency regulation, and load balancing through reasonable charge and discharge schedules [21]. As a vital part of the electrification of transportation, EVs can also be considered dynamic electricity storage batteries for the building community, in addition to transportation tasks [22,23]. Specifically, EVs can be charged with surplus renewable energy to power buildings when renewable energy generation is in short supply [24–26]. In addition, charging and discharging schedules and locations can be adjusted based on user demand and electricity price signals to reduce energy costs. Therefore, the integration of EVs into the building community can open possibilities for energy efficiency improvement and a low-carbon transition in the building sector.

4.1.2 Building energy flexibility

Several studies have proposed various definitions of building energy flexibility [27–29]. Despite the variety of quantitative indicators, it is generally believed that building energy flexibility refers to the ability of buildings to flexibly adjust their energy demand and generation according to local conditions, such as weather, to support the grid without compromising the production and comfort requirements of building occupants [30].

In the aspect of DSM, building energy flexibility is realized by altering the energy demand profile and on-site renewable energy generation [31], such as flexible end users (i.e., facilities that can adjust the demand curve [32]), and associated facilities that assist end users in adjusting their demand profiles [33,34]. According to Wang et al. [30], heat pumps, air conditioners, washers and dryers, and dishwashers are the end-use facilities that receive the most attention for demand-side flexibility [35,36]. In addition, building thermal mass, water tanks, batteries, PVs, etc., are the most commonly associated facilities that can be used to enhance flexible building operations [37−39]. In this chapter the building energy flexibility sources and technologies are further discussed in terms of thermostatically controlled load (TCL) regulation, power storage, and appliance dispatching.

TCL regulation assisted by thermal energy storage (TES) is the most common building flexibility technology [30], which can achieve up to 90% peak load drop shedding flexibility [40]. For the passive type, the building envelope and internal thermal mass can help store cooling or heating energy by precooling or preheating the building during off-peak hours, which helps shift the peak energy demand from HVAC systems [41,42]. As for the active type, TES tanks allow flexibility in load shifting and shedding by storing cooling or heating energy during off-peak hours and releasing it during peak hours [41,42]. For a more detailed TES analysis, please refer to [43].

Power storage, including static batteries for buildings and dynamic batteries for EVs, is another source of flexibility. It can not only charge to fill the valley when the renewable energy supply is lower than the energy demand but also discharge and cut peaks when the renewable energy supply exceeds the energy demand [44−46]. Battery-based control strategies are a vital technical solution for the energy management of buildings equipped with on-site renewable energy generation [30]. Batteries have been applied in numerous studies to promote the utilization and self-consumption rate of rooftop PVs and BIPV systems [47,48], and on-site wind turbines [49,50], where more than 50% of energy demand was met by solar and wind energy generation being economically viable.

Appliance dispatching refers to switching the operation of home appliances from a peak hour period to an off-peak hour period within acceptable time intervals for users [51]. The most common home appliances that offer flexibility are washers, dryers, and dishwashers. However, due to the uncontrollable behaviors of building occupants, this flexible resource has certain inherent uncertainties [52].

4.1.3 Interactions among buildings, electrical vehicles, and the grid

The energy interactions between buildings and EVs, including building-to-vehicle (B2V) and vehicle-to-building (V2B), can be achieved by carrying out a unit (i.e., a V2G—V2B unit) that controls bidirectional power flow to reduce the dependence of household usage and transportation on the grid [53]. Fig. 4.2 shows the diagram of the V2B interaction. According to the summary in Zhou et al. [53], the objectives of V2B consist of the annual matching ability [55—57], annual net operating cost [58,59], annual net equivalent CO_2 emission [60—62], and grid energy interaction [63,64].

In addition to battery-based EVs, building fuel cell EV (FCEV) energy interactions has received increasing attention in recent years with hydrogen storage and fuel cell technology breakthroughs [65—68]. The building FCEV energy interactions refer to the hydrogen storage tank vented to activate the fuel cell, which supports electricity for the normal and stable operation of the building. Hydrogen can be sourced either from hydrogen stations or from electrolyzers powered by renewable energy generation [53]. Energy integration between FCEVs and buildings lies in electrolyzer and fuel cell efficiencies, dynamic building loads, renewable energy capacity, uncertain driving schedules, etc. [53].

To achieve a real-time energy balance of the system, the power grid should cover the energy demand of buildings and EVs during renewable energy shortage periods and store the surplus renewable energy in the surplus period. Energy interaction with the grid depends on real-time building energy demand, renewable energy generation, charging and discharging of energy storage systems, etc. With the integration of EVs, energy consumption in transportation and energy losses in charging/discharging should be noted. In addition, the inherent intermittency and uncertainty of renewable energy generation will impose challenges on energy interaction networks.

4.2 Demand-side management

The stable and reliable operation of the power grid is highly dependent on the real-time energy balance between energy demand and power

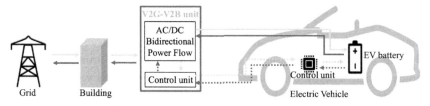

Figure 4.2 Diagram of vehicle-to-building energy interactions [53,54].

supply [69]. Assuming that only the supply side is controlled, the energy balance is difficult to maintain, especially with the increase in the proportion of renewable energy generation. Since renewable energy generation is highly dependent on weather conditions, it is hard to follow specific load shapes. Compared with the conventional method of providing corresponding electricity according to energy demand, DSM strategies have received increasing attention.

DSM can be summarized as the planning, implementation, and monitoring of utility activities aimed at influencing or changing the electricity demand of customers [70]. Consequently, the time pattern and magnitude of the utility's load can be changed. Commonly, DSM is used to encourage consumers to restrict electricity consumption behavior during peak hours and thereafter reduce energy consumption, or to shift electricity usage time to off-peak periods to flatten the energy demand profile [54]. Sometimes, it is desirable to make the curve follow the generation pattern. Both approaches to DSM require regulation of customers' energy use [70].

4.2.1 Load shaping techniques

DSM has been gradually investigated since the end of the 20th century [71,72]. DSM can be applied to complete various load shaping tasks, such as strategic conservation, strategic load growth, flexible load shape, peak clipping, valley filling, and load shifting, as shown in Fig. 4.3 [71]. The combination of these technologies can effectively facilitate the matching of load curve shape and power generation, to reduce the grid power supply, improve the utilization efficiency of renewable energy, and enhance the energy autonomy of the building EV system [73].

DSM strategies are categorized into five types [30], including load efficiency, load shedding, load shifting, load modulation, and load generation. Load efficiency refers to the continuous reduction in energy demand, which is a long-term reduction strategy in sustainable energy consumption. Load shedding refers to the reduction of short-term electricity demand during peak hours or emergencies. In this strategy, power demand is required to be reduced within a few minutes, and this reduction typically lasts for up to an hour. Load shifting specifically refers to the shifting of a building's energy usage time from peak hours to off-peak hours, where energy demand is requested to be reduced within minutes and this process typically lasts about two to four hours. Load modulation means that buildings automatically adjust their energy demand after

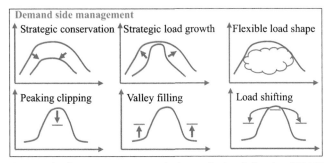

Figure 4.3 Demand-side management techniques for load shaping [70,71].

receiving a signal from the grid operator. In the load modulation strategy, the power demand will be adjusted on a second or subsecond basis. The generation refers to the on-site renewable energy generation system supplying the building with electricity during peak hours, where the energy demand is usually reduced within minutes, and this reduction typically lasts about two to four hours.

4.2.2 Electric energy storage technologies and electric vehicles

Electricity needs to be consumed or converted to be stored in another form when it is produced [74]. Shaping the load curve to follow renewable generation is nearly impossible due to the uncertainty and dynamics of natural resources, such as solar and wind. Energy storage technology buffers the instantaneous relationship between renewable energy and energy consumption to a certain extent and contributes to energy balance. Therefore energy storage technology is necessary for renewable energy utilization and DSM.

Advancements in battery storage technology could boost the number of EVs in cities. This will cause an increase in electricity consumption, especially during night-time peak periods [75]. Meanwhile, thousands of EVs could work as large-scale distributed energy storage systems with potential [76]. Energy integration, as well as interactions between EVs and buildings, is an effective strategy to support the energy demand of e-mobility with on-site renewable energy generation and distributed storage of surplus energy. For example, grid-connected buildings [77] are integrated with EVs, helping to provide energy to the buildings if solar energy generation is in short supply. Simulation results validate the energy flexibility and economic feasibility of the integration under appropriate control strategies. Besides, a smart charging and discharging process for multiple plug-in hybrid EVs integrated with a building has been proposed [78] to help decrease the peak power load and total energy cost based on a novel distributed algorithm.

4.2.3 Demand response

Flexibility enables building energy demand to be controlled and adjusted in response to weather conditions, grid requirements, or electricity price signals. The adjustment of power on the demand side is referred to as demand response (DR).

For external signals, such as changes in electricity prices, building energy users can generally respond in three ways [70]. In the first way, users can reduce the usage of electrical equipment only during peak hours, while the power consumption pattern is unnecessary to be adjusted during off-peak hours [79]. However, this response strategy results in compromised comfort for energy users since they are compelled to curb power demand for a specified period. In this way, total energy consumption and electricity bills are reduced. The second way lies in shifting power usage from peak hours to off-peak hours [80]. This approach can contribute to a flatter load curve by reducing peak loads and filling valleys. In this way, the total energy consumption of end users is not reduced, while the stable operation of the power system is achieved and the efficiency of power transmission and distribution is improved. The third way is to use on-site renewable energy generation to reduce dependence on the grid [81]. This method helps increase end-user energy autonomy, facilitates distributed generation, and reduces operational stress on distribution and transmission grids.

DR is technically challenging. Most electricity demand management technologies depend on high-bandwidth and reliable connections for establishing two-way communication between consumers and utilities and transmitting signals. [82]. In addition, DR is hardware-intensive, which means that the energy management equipment and intelligent household electrical appliances should be equipped to respond to price signals, etc., promptly without human intervention [70].

4.3 Model predictive control
4.3.1 Why model predictive control?

Model predictive control (MPC) establishes and solves a suboptimization problem at each sampling interval to determine the optimal operation plan subject to technical and physical constraints. Several advantages make MPC suitable for energy management and control of the integrated building e-mobility system. Specifically, the feedback mechanism in MPC allows the system to deal with uncertainties. Besides, physical constraints (such as generation, storage capacity, and demand forecasts) can be considered. In addition, MPC is highly dependent on the system's future behaviors, and accurate prediction will dominate the integrated building e-mobility systems under dynamic demand and renewable energy generation profiles [83]. The inherent closed-loop characteristic of MPC leads to quickly modifying the operation plan based on the system response, and the measured values are used as feedback to update the optimization problem. MPC has been studied for energy management tasks in microgrids with EV charging stations [84].

4.3.2 Model predictive control and its applications

MPC is not a specific control algorithm, yet it refers to a class of methods that have both predictive and control parts. These methods are based on system models to compute control signals by minimizing cost functions. As shown in Fig. 4.4, the mechanism of MPC is that at each sampling interval, an optimization problem, which is finite-time and open-loop, is solved online based on current measurement data. The first portion of the gained control sequence is then acted on the targeted object. At the next time interval, the above procedure is repeated: updated measurements are used as initial conditions to predict system dynamics in the future. Afterward, the optimization model is tuned and solved again. In short, the MPC algorithm includes three steps: (1) predicting the future performance of the targeted system; (2) numerically solving the open-loop optimization problem; and (3) employing the first portion of the optimal solution to

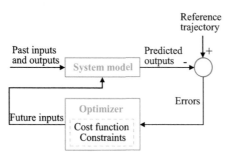

Figure 4.4 Schematic diagram of model predictive control algorithm [83].

the targeted system. The three steps are repeated at each sampling interval, and the newly measured value at each sampling interval is used as the initial condition for predicting the future performance of the system.

Three MPC approaches are used for building energy management, as shown in Fig. 4.5. Methods based on physical models are commonly referred to as white-box methods. White-box MPC relies on thermodynamic and heat transfer mechanisms to describe the energy performance of buildings. Data-driven approaches, that is, black-box approaches, mainly adopt statistical models or machine learning methods to assess the energy performance of buildings. A hybrid approach, or gray-box approach, refers to a combination of white-box and black-box models. For white-box and gray-box models, the most prominent simulation software is MATLAB®, TRNSYS, and EnergyPlus. The most commonly used tools for black-box models are Python and R [10].

MPC has been widely applied in the energy management of integrated energy systems with buildings and EVs. For example, a campus-integrated microgrid has been designed [85], which consists of PV parking shades, an energy storage device, two types of EVs, an advanced metering system, and a control module. A white-box MPC algorithm is used to reduce the peak load of the campus while satisfying the normal charging and operation of various EVs and energy storage systems. Results show that the MPC strategy can reduce the peak load under practical operational conditions by up to 15%. Besides, a Monte Carlo uncertainty analysis method has been adopted [86] to evaluate the robustness of the MPC method against the uncertainty of the EVs scheduling. The simulation results show that the MPC method is robust in comfort performance, where it is capable of maintaining temperature deviations below 1°C within 95% of the simulation time. In addition, a load aggregation method in a power distribution system based on MPC has been proposed [87], with the aim of aggregation control of the energy load of a large building complex. The results showed a 21% reduction in generation costs and a 17% reduction in peak load.

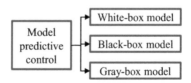

Figure 4.5 Classification of model predictive control approaches.

4.4 Challenges and future opportunities

4.4.1 Challenges

The integration of EVs and building energy systems is considered a promising technology for promoting renewable energy utilization and energy sharing. This integration can help increase the energy flexibility of buildings and reduce carbon emissions of buildings and transportation. However, to make the integration between EVs and building energy systems more promising and widely accepted, some technical and economic challenges still need to be resolved:

1. There is a lack of approaches and indicators to describe and quantify the impact of building energy flexible operations on EV mobility. As the number and proportion of private EVs increase, their impact on building energy and the impact of energy interaction requirements on their mobility are increasing. Such quantitative methods are needed to help evaluate the quantitative relationship between the energy interaction function of EVs and their mobility transport functions.

2. In the integrated energy-sharing system of EVs, without economic incentives for the depreciation of EV batteries, owners may be reluctant to perform additional charging to slow down the aging of batteries.

3. Due to the wide variety of commercial EVs on the market, the compatibility of different brands of EVs with building energy systems, and the timeliness and reliability of EVs for powering buildings should be further considered.

4. With the increase in the number of EVs and EV charging piles, the power grid may experience instability and fluctuations in the energy interaction between EVs and buildings. This can be a challenge for grid capacity and energy management strategies for district energy systems.

5. For model-based MPC approaches, accurate characterization of the performance of integrated building systems, for example, techno-economic-environmental performance, is fundamental in ensuring acceptable smart building control and automation execution.

6. Compared with traditional control methods, the MPC method is superior in control accuracy and stability, yet the computational efficiency should be further improved. Since there is a suboptimization problem for each interval, multiple iterative optimizations lead to a large computational load, especially with constraints and long horizons.

4.4.2 Future opportunities

Based on techno-economic challenges, future research opportunities are summarized as follows:

1. Approaches and indicators are needed to describe and quantify the impact of building energy flexible operations on EV mobility.
2. Research on economic incentive policies for battery depreciation of EVs to encourage the owners to charge additionally and support energy interaction and energy sharing with buildings.
3. The need to define and establish standardized building-EV interfaces for different commercial markets and models of EVs, as well as standards for B2V and V2B interactions.
4. Expand the capacity of the local grid and update the energy management strategy of the local energy system to support the electricity demand of a large number of EVs and supporting infrastructure.
5. The performance description formula of the integrated building-e-mobility system should be established accurately and even calibrated with real experimental data. Techniques (such as machine learning surrogate models) can be used to improve model accuracy and system performance through computational efficiency. In addition, artificial intelligence technologies (such as intelligent algorithms) can be beneficial to reduce optimization iteration time.

References

[1] Berardi U. A cross-country comparison of the building energy consumptions and their trends. Resources, Conservation and Recycling 2017;123:230−41.
[2] Nejat P, Jomehzadeh F, Taheri MM, Gohari M, Majid MZA. A global review of energy consumption, CO_2 emissions and policy in the residential sector (with an overview of the top ten CO_2 emitting countries). Renewable and Sustainable Energy Reviews 2015;43:843−62.
[3] Homod RZ. Analysis and optimization of HVAC control systems based on energy and performance considerations for smart buildings. Renewable Energy 2018;126: 49−64.
[4] Li H, Hong T, Lee SH, Sofos M. System-level key performance indicators for building performance evaluation. Energy and Buildings 2020;209:109703.
[5] Lumpkin DR, Horton WT, Sinfield JV. Holistic synergy analysis for building subsystem performance and innovation opportunities. Building and Environment 2020;178:106908.
[6] Aelenei D, Lopes RA, Aelenei L, Gonçalves H. Investigating the potential for energy flexibility in an office building with a vertical BIPV and a PV roof system. Renewable Energy 2019;137:189−97.
[7] Chen T, Tai KF, Raharjo GP, Heng CK, Leow SW. A novel design approach to prefabricated BIPV walls for multi-storey buildings. Journal of Building Engineering 2023;63:105469.

[8] Ourahou M, Ayrir W, Hassouni BE, Haddi A. Review on smart grid control and reliability in presence of renewable energies: challenges and prospects. Mathematics and Computers in Simulation 2020;167:19−31.

[9] Lu F, Yu Z, Zou Y, Yang X. Energy flexibility assessment of a zero-energy office building with building thermal mass in short-term demand-side management. Journal of Building Engineering 2022;50:104214.

[10] Mariano-Hernández D, Hernández-Callejo L, Zorita-Lamadrid A, Duque-Pérez O, García FS. A review of strategies for building energy management system: model predictive control, demand side management, optimization, and fault detect & diagnosis. Journal of Building Engineering 2021;33:101692.

[11] Chen Y, Xu P, Gu J, Schmidt F, Li W. Measures to improve energy demand flexibility in buildings for demand response (DR): a review. Energy and Buildings 2018;177:125−39.

[12] Tang H, Wang S. Energy flexibility quantification of grid-responsive buildings: energy flexibility index and assessment of their effectiveness for applications. Energy. 2021;221:119756.

[13] Zhou Y, Cao S. Energy flexibility investigation of advanced grid-responsive energy control strategies with the static battery and electric vehicles: a case study of a high-rise office building in Hong Kong. Energy Conversion and Management 2019;199: 111888.

[14] Zhou Y. Energy sharing and trading on a novel spatiotemporal energy network in Guangdong-Hong Kong-Macao Greater Bay Area. Applied Energy 2022;318:119131.

[15] Zhou Y, Cao S, Hensen JL. An energy paradigm transition framework from negative towards positive district energy sharing networks—battery cycling aging, advanced battery management strategies, flexible vehicles-to-buildings interactions, uncertainty and sensitivity analysis. Applied Energy 2021;288:116606.

[16] Zhou Y, Cao S, Hensen JL, Hasan A. Heuristic battery-protective strategy for energy management of an interactive renewables−buildings−vehicles energy sharing network with high energy flexibility. Energy Conversion and Management 2020;214: 112891.

[17] Zhou Y, Cao S, Kosonen R, Hamdy M. Multi-objective optimisation of an interactive buildings-vehicles energy sharing network with high energy flexibility using the Pareto archive NSGA-II algorithm. Energy Conversion and Management 2020;218: 113017.

[18] Zafar R, Mahmood A, Razzaq S, Ali W, Naeem U, Shehzad K. Prosumer based energy management and sharing in smart grid. Renewable and Sustainable Energy Reviews 2018;82:1675−84.

[19] Mahmood D, Javaid N, Ahmed G, Khan S, Monteiro V. A review on optimization strategies integrating renewable energy sources focusing uncertainty factor−Paving path to eco-friendly smart cities. Sustainable Computing: Informatics and Systems 2021;30:100559.

[20] Xu Z, Guan X, Jia Q-S, Wu J, Wang D, Chen S. Performance analysis and comparison on energy storage devices for smart building energy management. IEEE Transactions on Smart Grid 2012;3:2136−47.

[21] Cui T, Chen S, Wang Y, Zhu Q, Nazarian S, Pedram M. An optimal energy co-scheduling framework for smart buildings. Integration (Tokyo, Japan) 2017;58: 528−37.

[22] Gong H, Ionel DM. Improving the power outage resilience of buildings with solar PV through the use of battery systems and EV energy storage. Energies. 2021;14: 5749.

[23] Gough R, Dickerson C, Rowley P, Walsh C. Vehicle-to-grid feasibility: a techno-economic analysis of EV-based energy storage. Applied Energy 2017;192:12−23.

[24] Barone G, Buonomano A, Forzano C, Giuzio GF, Palombo A. Increasing self-consumption of renewable energy through the building to vehicle to building approach applied to multiple users connected in a virtual micro-grid. Renewable Energy 2020;159:1165−76.

[25] He Z, Khazaei J, Freihaut JD. Optimal integration of vehicle to building (V2B) and building to vehicle (B2V) technologies for commercial buildings. Sustainable Energy, Grids and Networks 2022;32:100921.

[26] Huang P, Lovati M, Zhang X, Bales C. A coordinated control to improve performance for a building cluster with energy storage, electric vehicles, and energy sharing considered. Applied Energy 2020;268:114983.

[27] Bampoulas A, Saffari M, Pallonetto F, Mangina E, Finn DP. A fundamental unified framework to quantify and characterise energy flexibility of residential buildings with multiple electrical and thermal energy systems. Applied Energy 2021;282:116096.

[28] Luc KM, Heller A, Rode C. Energy demand flexibility in buildings and district heating systems—a literature review. Advances in Building Energy Research 2019;13: 241−63.

[29] Zhou Y, Cao S. Quantification of energy flexibility of residential net-zero-energy buildings involved with dynamic operations of hybrid energy storages and diversified energy conversion strategies. Sustainable Energy, Grids and Networks 2020;21:100304.

[30] Li H, Wang Z, Hong T, Piette MA. Energy flexibility of residential buildings: a systematic review of characterization and quantification methods and applications. Advances in Applied Energy 2021;3:100054.

[31] Chen L, Xu Q, Yang Y, Song J. Optimal energy management of smart building for peak shaving considering multi-energy flexibility measures. Energy and Buildings 2021;241:110932.

[32] Tulabing RS, Mitchell BC, Covic GA, Boys JT. Localized demand control of flexible devices for peak load management. IEEE Transactions on Smart Grid 2022;14:217−27.

[33] Wijesuriya S, Booten C, Bianchi MV, Kishore RA. Building energy efficiency and load flexibility optimization using phase change materials under futuristic grid scenario. Journal of Cleaner Production 2022;339:130561.

[34] Zhou Y. Demand response flexibility with synergies on passive PCM walls, BIPVs, and active air-conditioning system in a subtropical climate. Renewable Energy 2022;199:204−25.

[35] Chen Y, Chen Z, Xu P, Li W, Sha H, Yang Z, et al. Quantification of electricity flexibility in demand response: office building case study. Energy. 2019;188:116054.

[36] Ren H, Sun Y, Albdoor AK, Tyagi V, Pandey A, Ma Z. Improving energy flexibility of a net-zero energy house using a solar-assisted air conditioning system with thermal energy storage and demand-side management. Applied Energy 2021;285: 116433.

[37] Chen Y, Xu P, Chen Z, Wang H, Sha H, Ji Y, et al. Experimental investigation of demand response potential of buildings: combined passive thermal mass and active storage. Applied Energy 2020;280:115956.

[38] Dias JB, da Graça GC. Using building thermal mass energy storage to offset temporary BIPV output reductions due to passing clouds in an office building. Building and Environment 2022;207:108504.

[39] Rodríguez LR, Ramos JS, Delgado MG, Félix JLM, Domínguez SÁ. Mitigating energy poverty: potential contributions of combining PV and building thermal mass storage in low-income households. Energy Conversion and Management 2018;173: 65−80.

[40] Sehar F, Pipattanasomporn M, Rahman S. An energy management model to study energy and peak power savings from PV and storage in demand responsive buildings. Applied Energy 2016;173:406−17.

[41] Johra H, Heiselberg P, Le, Dréau J. Influence of envelope, structural thermal mass and indoor content on the building heating energy flexibility. Energy and Buildings 2019;183:325−39.

[42] Le Dréau J, Heiselberg P. Energy flexibility of residential buildings using short term heat storage in the thermal mass. Energy. 2016;111:991−1002.

[43] Finck C, Li R, Kramer R, Zeiler W. Quantifying demand flexibility of power-to-heat and thermal energy storage in the control of building heating systems. Applied Energy 2018;209:409−25.

[44] Li N, Uckun C, Constantinescu EM, Birge JR, Hedman KW, Botterud A. Flexible operation of batteries in power system scheduling with renewable energy. IEEE Transactions on Sustainable Energy 2015;7:685−96.

[45] Liu W, Song MS, Kong B, Cui Y. Flexible and stretchable energy storage: recent advances and future perspectives. Advanced Materials 2017;29:1603436.

[46] Stavrakas V, Flamos A. A modular high-resolution demand-side management model to quantify benefits of demand-flexibility in the residential sector. Energy Conversion and Management 2020;205:112339.

[47] Rababah HE, Ghazali A, Mohd Isa MH. Building integrated photovoltaic (BIPV) in Southeast Asian countries: Review of effects and challenges. Sustainability. 2021;13:12952.

[48] Sharma P, Kolhe M, Sharma A. Economic performance assessment of building integrated photovoltaic system with battery energy storage under grid constraints. Renewable Energy 2020;145:1901−9.

[49] González-Mahecha RE, Lucena AF, Szklo A, Ferreira P, Vaz AIF. Optimization model for evaluating on-site renewable technologies with storage in zero/nearly zero energy buildings. Energy and Buildings 2018;172:505−16.

[50] Liu J, Chen X, Yang H, Shan K. Hybrid renewable energy applications in zero-energy buildings and communities integrating battery and hydrogen vehicle storage. Applied Energy 2021;290:116733.

[51] Jindal A, Bhambhu BS, Singh M, Kumar N, Naik K. A heuristic-based appliance scheduling scheme for smart homes. IEEE Transactions on Industrial Informatics 2019;16:3242−55.

[52] Sadat-Mohammadi M, Nazari-Heris M, Nazerfard E, Abedi M, Asadi S, Jebelli H. Intelligent approach for residential load scheduling. IET Generation, Transmission & Distribution 2020;14:4738−45.

[53] Zhou Y, Cao S, Hensen JL, Lund PD. Energy integration and interaction between buildings and vehicles: a state-of-the-art review. Renewable and Sustainable Energy Reviews 2019;114:109337.

[54] Ioakimidis CS, Thomas D, Rycerski P, Genikomsakis KN. Peak shaving and valley filling of power consumption profile in non-residential buildings using an electric vehicle parking lot. Energy. 2018;148:148−58.

[55] Cao S, Hasan A, Sirén K. On-site energy matching indices for buildings with energy conversion, storage and hybrid grid connections. Energy and Buildings 2013;64:423−38.

[56] Jensen SØ, Marszal-Pomianowska A, Lollini R, Pasut W, Knotzer A, Engelmann P, et al. IEA EBC annex 67 energy flexible buildings. Energy and Buildings 2017;155:25−34.

[57] Reynders G, Lopes RA, Marszal-Pomianowska A, Aelenei D, Martins J, Saelens D. Energy flexible buildings: an evaluation of definitions and quantification methodologies applied to thermal storage. Energy and Buildings 2018;166:372−90.

[58] Amirioun MH, Kazemi A. A new model based on optimal scheduling of combined energy exchange modes for aggregation of electric vehicles in a residential complex. Energy. 2014;69:186−98.

[59] Fang Y, Asche F, Novan K. The costs of charging plug-in electric vehicles (PEVs): Within day variation in emissions and electricity prices. Energy Economics 2018;69: 196—203.

[60] Hoehne CG, Chester MV. Optimizing plug-in electric vehicle and vehicle-to-grid charge scheduling to minimize carbon emissions. Energy. 2016;115:646—57.

[61] Karan E, Mohammadpour A, Asadi S. Integrating building and transportation energy use to design a comprehensive greenhouse gas mitigation strategy. Applied Energy 2016;165:234—43.

[62] Fernández RÁ. A more realistic approach to electric vehicle contribution to greenhouse gas emissions in the city. Journal of Cleaner Production 2018;172:949—59.

[63] Verbruggen B, Driesen J. Grid impact indicators for active building simulations. IEEE Transactions on Sustainable Energy 2014;6:43—50.

[64] Voss K., Sartori I., Napolitano A., Geier S., Goncalves H., Hall M., et al. Load matching and grid interaction of net zero energy buildings. EUROSUN 2010 international conference on solar heating, cooling and buildings. 2010.

[65] Strubel V. STORHY: hydrogen storage systems for automotive applications. Magna Steyr Fahrzeugtechnik AG&Co KG; 2008.

[66] Hwang JJ. Review on development and demonstration of hydrogen fuel cell scooters. Renewable and Sustainable Energy Reviews 2012;16:3803—15.

[67] Zhang L, Xiang J. The performance of a grid-tied microgrid with hydrogen storage and a hydrogen fuel cell stack. Energy Conversion and Management 2014;87:421—7.

[68] Ramirez JPB, Halm D, Grandidier J-C, Villalonga S, Nony F. 700 bar type IV high pressure hydrogen storage vessel burst—Simulation and experimental validation. International Journal of Hydrogen Energy 2015;40:13183—92.

[69] Rathor SK, Saxena D. Energy management system for smart grid: an overview and key issues. International Journal of Energy Research 2020;44:4067—109.

[70] Gelazanskas L, Gamage KA. Demand side management in smart grid: a review and proposals for future direction. Sustainable Cities and Society 2014;11:22—30.

[71] Gellings CW. The concept of demand-side management for electric utilities. Proceedings of the IEEE 1985;73:1468—70.

[72] Gellings C.W., Chamberlin J.H. Demand-side management: concepts and methods. 1987.

[73] Pina A, Silva C, Ferrão P. The impact of demand side management strategies in the penetration of renewable electricity. Energy. 2012;41:128—37.

[74] Bhat I, Prakash R. LCA of renewable energy for electricity generation systems—a review. Renewable and Sustainable Energy Reviews 2009;13:1067—73.

[75] Muratori M. Impact of uncoordinated plug-in electric vehicle charging on residential power demand. Nature Energy 2018;3:193—201.

[76] Alshahrani S, Khalid M, Almuhaini M. Electric vehicles beyond energy storage and modern power networks: challenges and applications. IEEE Access 2019;7:99031—64.

[77] Korkas CD, Terzopoulos M, Tsaknakis C, Kosmatopoulos EB. Nearly optimal demand side management for energy, thermal, EV and storage loads: an approximate dynamic programming approach for smarter buildings. Energy and Buildings 2022;255:111676.

[78] Nguyen HK, Song JB. Optimal charging and discharging for multiple PHEVs with demand side management in vehicle-to-building. Journal of Communications and Networks 2012;14:662—71.

[79] Dagdougui H, Ouammi A, Dessaint LA. Peak load reduction in a smart building integrating microgrid and V2B-based demand response scheme. IEEE Systems Journal 2018;13:3274—82.

[80] Klein K, Herkel S, Henning H-M, Felsmann C. Load shifting using the heating and cooling system of an office building: quantitative potential evaluation for different flexibility and storage options. Applied Energy 2017;203:917—37.

[81] Cao X, Zhang J, Poor HV. Data center demand response with on-site renewable generation: A bargaining approach. IEEE/ACM Transactions on Networking 2018;26:2707−20.

[82] Poolo I. A smart grid demand side management framework based on advanced metering infrastructure. American Journal of Electrical and Electronics Engineering 2017;5:152−8.

[83] Bordons C, Garcia-Torres F, Ridao MA. Model predictive control of microgrids. Springer; 2020.

[84] Zhang Y, Wang R, Zhang T, Liu Y, Guo B. Model predictive control-based operation management for a residential microgrid with considering forecast uncertainties and demand response strategies. IET Generation, Transmission & Distribution 2016;10:2367−78.

[85] Achour Y, Ouammi A, Zejli D. Model predictive control based demand response scheme for peak demand reduction in a Smart Campus Integrated Microgrid. IEEE Access 2021;9:162765−78.

[86] Seal S, Boulet B, Dehkordi VR, Bouffard F, Joos G. Centralized MPC for home energy management with EV as mobile energy storage unit. IEEE Transactions on Sustainable Energy 2023;.

[87] Mirakhorli A, Dong B. Model predictive control for building loads connected with a residential distribution grid. Applied Energy 2018;230:627−42.

CHAPTER 5

Electrification and hydrogenation in integrated building-transportation systems for sustainability

Bin Gao
Sustainable Energy and Environment Thrust, Function Hub, The Hong Kong University of Science and Technology (Guangzhou), Nansha, Guangdong, P.R. China

5.1 Introduction

Electrification and hydrogenation technologies have been widely used in industries and public life for the purpose of carbon neutrality. To decarbonize both buildings and transportation sectors, electrochemical battery storage and hydrogen energy storage have been adopted in vehicles with power supply sources, community energy supply, and grid stabilization, owing to their environment-friendly characteristics. Electrochemical battery storage can store electricity and discharge it for demand coverage. The battery is generally used for short-term energy storage because of its fast response. Hydrogen energy storage supplies hydrogen to generate power and heat through fuel cells with by-product water only. Hydrogen can be used for the purpose of long-term energy storage, such as seasonal and annual storage, because of the high gravimetric density and stability of the storage tanks.

Currently, battery devices and hydrogen devices have undergone rapid development in vehicles, such as battery electric vehicles (EVs) (BEVs) of BYD [1] and hydrogen fuel cell EVs (FCEVs) of Toyota [2]. Electrification and hydrogenation in buildings, including stationary energy storage integration [3–6] and mobile energy storage integration by vehicles, are attracting the interest of researchers and engineers [7–12]. Nowadays, studies mostly focus on energy supply [13–15], transportation applications [16–18], and small-scale energy interaction in integrated building and transportation systems [9,19]. However, a systematic review of electrification and hydrogenation techniques is quite limited. Functions and mechanisms of electrification and hydrogenation are not clear in

Advances in Digitalization and Machine Learning for Integrated
Building-Transportation Energy Systems
DOI: https://doi.org/10.1016/B978-0-443-13177-6.00009-6

integrated multienergy systems, as well as in newly interactive energy frameworks and technical challenges.

This study presents a holistic and systematic review to discuss electrification and hydrogenation in integrated building-transportation systems. The techniques and applications involved in building transportation systems of electrochemical battery storage and hydrogen storage have been introduced. The underlying mechanism and role of electrification and hydrogenation in building-transportation systems were characterized, including stationary storages, mobile storages, decentralized and centralized battery storages, and integration of micro-hydrogen systems and hydrogen stations with buildings. Finally, technical challenges and low-carbon community energy frameworks have been discussed to promote district decarbonization.

5.2 Electrification techniques

5.2.1 Electrochemical battery storages

Electrochemical battery storages with fast power response have been widely applied in industries for a long time. Based on the electrochemical materials for storing energy, batteries can be divided into lithium-ion batteries, lead acid batteries, nickel-metal hydride batteries, nickel-cadmium batteries, sodium-sulfur batteries, sodium nickel chloride batteries, etc. [20].

Compared with other battery types, lithium-ion batteries possess high power density (75−200 Wh/kg), energy efficiency (higher than 90%), and lifetime (10,000 cycles at 100% discharge depth) [21]. Lithium-ion batteries have been widely used in transportations, buildings, and electric grids. Lead acid batteries with secure and stable characteristics are used as power sources in heavy vehicles for uninterruptible power supply and spinning reserve. However, lead acid batteries have a much lower power density (30−50 Wh/kg) and lifetime (200−1000 cycles at 100% discharge depth) compared with lithium-ion batteries [22] which greatly limits their widespread applications. Nickel-metal hydride batteries have attracted attention in EVs and hybrid EVs because of the advantages of flexible cell sizes, high operation voltage, flexible vehicle packaging, excellent thermal properties, environmental friendliness, etc. [23]. However, nickel-metal hydride batteries are difficult to fast charge and overcharge. The nickel-cadmium battery is one of the most popular battery types because of its superior advantages in low internal resistance, stable performance at low temperatures, and fast recharge [24]. However, because of cadmium toxicity, nickel-cadmium batteries are regarded as unfriendly to the environment, which results in their application prohibition in some European countries. Sodium-sulfur batteries need to operate at a temperature of 300°C for the electrode materials to be kept in a molten conducting state, and they are capable of reaching a high theoretical power density of 760 Wh/kg [15], even higher than that of lithium-ion batteries at 200 Wh/kg. The largest sodium-sulfur battery station was established in Presidio, Texas, with 4-MW power for up to 8 hours operation [20]. However, due to the possible mixture of molten sodium and sulfur during operation, the potential explosion crisis significantly restrains its commercial development. Sodium nickel chloride batteries have been increasingly adopted in power sources of EVs [25−28] due to their high energy efficiency (85%−95%) and durable lifetime (4500 cycles at 100% discharge depth) [29]. Similar to sodium-sulfur batteries, the sodium salt in sodium nickel chloride batteries also needs to stay at a melting state for electric conduction, so the optimum temperature of operation is in the range of 270°C−350°C, which is much higher than the melting point of 157°C. The reaction schematics of the battery types are characterized in Fig. 5.1. Batteries commonly consist of electrodes, electrolytes, and reactants. Reactants adhering to electrodes release electrons under redox reactions, and electricity is produced by electron release and transportation in circuits.

5.2.2 Electrification in buildings and transportations

Traditional ways of achieving electrification of district buildings include integrating electrochemical stationary batteries into buildings for supplying daily electricity. One of the most significant roles of stationary battery building integration is reducing electricity costs from electric grids. Pandžić [6] formulated a model to calculate the optimal power capacity of a stationary battery for the most economical electricity cost, and tested the model in a hotel in Croatia. Rosati et al. [5] evaluated the environmental and economic impact of stationary lithium-ion batteries in a residential building under different electric capacities and climates and found that electricity from battery storage could not completely fill the temporary usage shortage. Mehrtash [4] proposed an enhanced mixed-integer nonlinear programming model to analyze the optimal size of an energy storage system integrated with a battery, and the model could evaluate the environmental impact of buying electricity from the grid and the energy storage system. Sepúlveda-Mora [31] investigated the effect of grid electric

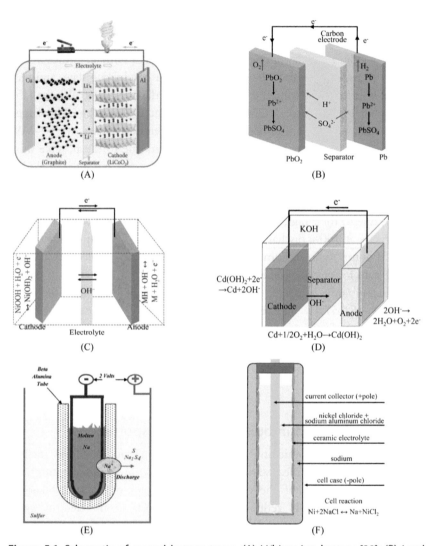

Figure 5.1 Schematic of several battery types. (A) Lithium-ion battery [30]; (B) Lead acid battery [20]; (C) Nickel-metal hydride battery; (D) Nickel-cadmium battery [24]; (E) Sodium-sulfur battery [15]; (F) Sodium nickel chloride battery [29].

tariff structure on the economic value of battery energy storage combined with buildings and pointed out that delaying the degradation limitations of battery replacement could bring substantial benefits.

Transportations account for a large proportion of carbon emissions due to the consumption of conventional fossil fuels. Electrification in transportation power sources is a hot issue in academia. Energy interactions between

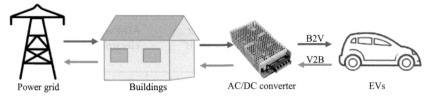

Figure 5.2 Schematic of V2B and B2V energy interactions. *B2V*, Building-to-vehicle; *V2B*, vehicle-to-building.

buildings and battery-based EVs have attracted widespread interest in recent years. The schematic of vehicle-to-building (V2B) and building-to-vehicle (B2V) energy interactions is shown in Fig. 5.2. The EV with a battery energy source is one of the most promising transportation types for reducing tailpipe emissions. Now a number of battery types are being adopted to equip the EVs, including lead acid, nickel–metal hydride, sodium nickel chloride, lithium-ion, etc. [18]. Due to its excellent energy efficiency, lifetime, and power density, the lithium-ion battery vehicle is the most popular EV type to be integrated with buildings [10−12,32].

5.3 Hydrogenation techniques

5.3.1 Hydrogen production

Hydrogen plays an increasingly significant role in carbon-neutral transition because of its advantages of high energy density, nonpollution, nontoxicity, and substantial reserves. In recent years, many studies have focused on the hydrogen production process to improve production efficiency and reduce carbon emissions. The feedstocks of hydrogen production can be classified into fossil fuels (crude oil, coal, and natural gas) and renewable resources (water and biomass), which can produce around 87 million tons of hydrogen per year [33]. Currently, fossil fuels occupy a high proportion of hydrogen production feedstocks, which could not be sustainable and inevitably generate carbon emissions. The development of the techniques of hydrogen feedstocks from water and biomass is a hot issue, and more than 60% of research works has been devoted to the investigation of water feedstocks [34].

Water electrolysis technologies for hydrogen production use electricity to split water molecules. With the development of electrolysis technologies, four types of ionic agents, which include solid oxide, proton exchange membrane (PEM), anion exchange membrane (AEM), and alkaline water

Table 5.1 Characteristics of electrolysis technology [33].

	Solid oxide water electrolysis	PEM water electrolysis	AEM water electrolysis	Alkaline water electrolysis
Operating temperature ($°C$)	700−850	50−80	40−60	70−90
Nominal current density (A/cm^2)	0.3−1	1−2	0.2−2	0.2−0.8
Cell voltage range (V)	1.0−1.5	1.4−2.5	1.4−2.0	1.4−3
Energy efficiency (%)	89	50−83	57−59	50−78
Lifetime (h)	20,000	50,000−80,000	> 30,000	60,000
Cell pressure (bar)	1	< 70	< 35	< 30

AEM, Anion exchange membrane; *PEM*, proton exchange membrane.

electrolysis, have emerged [33]. Table 5.1 shows the characteristics and comparison among the four electrolysis technologies. The solid oxide water electrolysis technology operates at a high temperature of 700°C−850°C and splits the water into oxygen and hydrogen at the vapor state. Although its energy efficiency is higher than the other types, its lifetime is limited because of the high-temperature operation conditions, resulting in larger energy consumption as well. PEM water electrolysis operates at a low-temperature condition and produces high-purity hydrogen of more than 99.999%; it also possesses a predominant lifetime of more than 50,000 working hours [35]. Thus the PEM water electrolysis technology is the most popular method of producing hydrogen. AEM water electrolysis is famous for its economic efficiency and low operating-temperature conditions. The most important limitation in commercial applications is their working life cycle. Kang et al. [36] used an AEM without an aryl-ether backbone structure to develop a stable AEM water electrolysis and verified the durability performance, which was shown to be better than conventional AEMs. Alkaline water electrolysis is a mature technology that has been developing for hundreds of years [37]. It adopts a concentrated alkaline solution, such as KOH/NaOH, as the electrolyte, and is also suitable for large-scale system applications.

5.3.2 Hydrogen energy storage

Hydrogen energy can be stored and migrated between areas to compensate for energy shortages in uneven spatiotemporal distribution. Unlike electricity stored in batteries with capacity degradation, there is no attenuation in hydrogen storage capacity, and hydrogen can be stored for a long period of time without any loss. To achieve a high volume-energy density, hydrogen storage is always kept at a high-pressure state. Thus the storage methods and materials need to meet strict safety requirements. Table 5.2 describes the classification of hydrogen storage.

Physical-based and material-based methods are employed in industrial practice for hydrogen storage. Compressed gas is usually stored in cylinders with a pressure of 100−700 atmospheres, and high-pressure hydrogen cylinders are most frequently used in industrial practice and lab studies. Liquid hydrogen can be achieved when the ambient temperature is lower than the boiling point at 1 atmosphere, that is, −252.8°C of hydrogen. Cold/cryo-compressed hydrogen storage has the advantages of high density and low power consumption [38]. However, the supply efficiency of hydrogen is limited because of the low temperature in hydrogen storage. As such the thermal management of cold/cryo-compressed hydrogen storage technology is necessary to improve hydrogen supply flexibility [39−41].

To improve material-based hydrogen storage technologies, the development of advanced materials has been dramatic in recent years. Larpruenrudee et al. [42] investigated the structure of metal hydride storage using numerical simulation and optimized the heat exchange efficiency. Sathe et al. [43] reviewed the development of high hydrogen storage capacity materials and the onboard reversibility of the host at

Table 5.2 Classification of hydrogen storage [21].

Classification principles	Storage methods
Physical-based	Compressed gas
	Liquid hydrogen
	Cold/Cryo compressed
Material-based	Interstitial hydride
	Absorbent
	Complex hydride
	Liquid organic
	Chemical hydrogen

operable thermodynamic conditions. They reported the critical material advancements of a systematic and sustainable hydrogen economy in this decade. Xu et al. [44] summarized the metal hydrogen storage materials, and the methods to improve the hydrogen storage performance of TiFe-based alloys. They proposed that enhancing the activity of ball-milled alloys and mixing transition metals and rare earth elements with high valence and affinity for hydrogen could optimize the storage performance of hydrogen. Sui et al. [45] pointed out the capacity shortage of magnesium-based hydrogen storage materials with nanostructures and reviewed the methods to improve the kinetic and thermodynamic properties of materials. Adding catalysts and alloys with other transition elements could improve hydrogen supply efficiency. Samimi et al. [46] synthesized different electrode materials based on zinc molybdate and tested their performance. They found that $Zn_3Mo_2O_9/ZnMoO_4$ and $Zn_3Mo_2O_9/ZnO$ possessed better hydrogen capacity than pure $Zn_3Mo_2O_9$.

5.3.3 Hydrogenation in buildings and transportations

A fuel cell is a kind of device that can convert chemical energy into electricity by consuming renewable energy. With the development of fuel cell technologies they came to be classified based on the electrolyte materials, such as proton exchange or polymer electrolyte membrane fuel cells (PEMFCs), solid oxide fuel cells, alkaline fuel cells, molten carbonate fuel cells, phosphoric acid fuel cells, etc. PEMFC is capable of converting hydrogen energy into electricity, and it is one of the most promising fuel cell technologies to replace conventional fossil devices because of its excellent advantages of high-power density, nonpollution, low noise, low operating temperature, and high economic benefits [47]. PEMFC operates at a temperature range of $50°C-90°C$, which needs to be maintained by liquid cooling or air cooling. Therefore liquid cooling PEMFC is able to supply not only electric energy but also heat energy using hot water, which could be used in domestic communities. Currently, a lot of research focuses on the PEMFC combined heat and power (CHP) cogeneration system in building applications. Fig. 5.3 describes the CHP cogeneration system in buildings. The heat and power supplies need to be balanced in accordance with the demand load in buildings. The energy management strategy is critical, which needs to match the CHP cogeneration system with other component operations for purposes of reducing the stack start times, saving energy usage [48], controlling energy efficiency [3], etc.

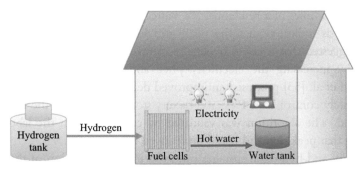

Figure 5.3 Applications of fuel cell CHP cogeneration systems in buildings. *CHP,* Combined heat and power.

In addition, migrating-storage hydrogen is always used in vehicles, and it is supplied to fuel cells as the electricity power source, that is, the FCEVs. Compared with BEVs, the advantages of FCEVs include longer driving endurance once charged, shorter charging time, more durable life cycles, and nonpollution in material recycling. Energy interactions between FCEVs and buildings combine the advantages of spatiotemporal energy migration and CHP cogeneration supply. To improve economic efficiency, energy flexibility, spatiotemporal energy balance, etc., energy management and migration strategies have been investigated [8,9,19,49−51]. However, the cost of FCEVs is higher than that of EVs and conventional vehicles, and their security is still controversial, which restricts the commercial development of FCEVs.

5.4 Functions and mechanisms of electrification and hydrogenation in buildings

The battery and hydrogen storage play critical roles in power energy systems, including peak shaving/valley filling, power stabilization, frequency regulation and spinning reserve, energy sharing in energy network, etc. [21].

5.4.1 Stationary battery and hydrogen storages

Peak shaving and valley filling for community electricity load are essential functions of battery and hydrogen storages. The surplus renewable energy is stored in storage systems under the load valley period, and discharged to cover demand under the demand peak period, shifting the electricity demand from peak to off-peak period. Applications of stationary storages and vehicle storages have been investigated. Li et al. [52] simulated the techno-economies of a residential PV-battery system connected to the grid. The analysis indicated that the peak grid load can be reduced by 1.1%. They suggested that it could benefit residential renewable system popularization by increasing the electricity price and reducing the battery cost. Mahmud et al. [53] proposed an improved decision-tree-based algorithm to coordinately control EVs, PVs, and battery storages to reduce the residential peak load. The algorithm was validated by experiments in the laboratory, showing that peak demand could be reduced by more than 9%.

Battery and hydrogen storages can assist in the interaction of the intermittent renewable energy with the grid to reduce power fluctuation. Furthermore, hybrid battery-hydrogen storages have the advantages of flexible response (interaction with the battery) and long-term storage (interaction with hydrogen), which benefit the stability of the power grid in both daily and long-term interactions. He et al. [8] proposed an energy management strategy to enhance energy flexibility and grid stability in a hybrid electricity-hydrogen sharing system. The results indicated that the strategy could reduce the maximum mean hourly grid power by 24.2% and the annual energy cost by 38.9%. Besides the energy management strategy, storage capacity plays a critical role, and oversizing electrolyzers could enhance the energy flexibility of the intermittent renewable energy and the grid [7].

Frequency regulation and spinning reserve for building devices and power grids are other critical functions of batteries and hydrogen storages. Since the load perturbations of power systems affects the grid frequency [54], it is necessary to reduce the frequency fluctuations and response required of the generating units. He et al. [55] analyzed a real-time cooperation scheme to integrate wind power and batteries with buildings for frequency regulation. They found that the scheme could improve wind regulation performance and increase overall revenue. Shi et al. [56] used a battery storage system for peak shaving and frequency regulation in a joint optimization framework. The experiment results indicated that this framework could reduce the electricity bills of users by 12%. Peng et al. [57] reviewed the dispatching strategies of EVs participating in frequency regulations and analyzed the stability and economy of such dispatching strategies.

Energy shifting is a cost-effective way of balancing energy expenses and consumption peaks in different time periods. Battery storages can realize load shifting in a short period of time because of its rapid charging and discharging characteristics [58]. Tang et al. [59] proposed dispatching strategies of thermal energy storage and power storage in buildings for full-scale flexibility, and the test results indicated that economic benefits could be increased by more than 13%. Hydrogen storages have been generally used for seasonal energy shifting because of their characteristic of long-term durable storage, and seasonal energy shifting can result in a low-carbon and low-priced electricity supply [60]. Kilic et al. [61] investigated the off-grid hybrid battery-hydrogen storage in daily and seasonal variation by simulation. The results showed that hybrid energy storages could reduce annual emissions by 68%−78% for battery storage and 84%−90% of hydrogen storage compared with a diesel energy system.

5.4.2 Mobile energy storages for sharing and migration

As far as battery and hydrogen storages in vehicles are concerned, they are capable of transporting and sharing electricity in a coordinated energy network, and it is effective in balancing the renewable energy in different areas by intercity energy transport frameworks, as shown in Fig. 5.4.

For battery storage applications, Zhou et al. [62] established a mathematical model to characterize the cycling aging of battery storage in multidirectional interactions of renewables−buildings−vehicles energy sharing network and developed an energy control strategy for battery life cycle protection. Zhou et al. [63] developed an interactive buildings−vehicles energy sharing network with multidirectional energy interactions and assessed several energy-related conflicts through multi-objective optimization. The results indicated that the equivalent CO_2 emissions were reduced by 7.5%. Besides, the techno-economic

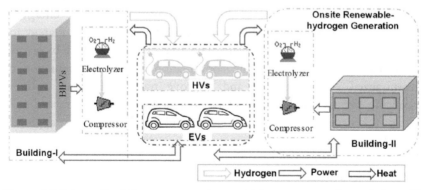

Figure 5.4 Schematic of the energy sharing and interarea energy transmission with BEVs and FCEVs [21]. *BEVs,* Battery electric vehicles; *FCEVs,* fuel cell electric vehicles.

performance of EVs—buildings interaction has been analyzed, such as net present value, discounted payback time, and net direct energy consumption [64].

For hydrogen storage applications, Robledo et al. [65] analyzed an FCEV-to-grid operation framework with a combined mobility and power generation and assessed its economic potential. The results showed that the V2G mode integrated with an FCEV could reduce the annual demand for electricity from the grid by 71%. He et al. [51] proposed an intercity transportation-based energy migration framework, and the renewable energy was transported in spatially different regions by FCEVs. The transportation-based energy migration framework covered 23.2% of office energy demand and increased the renewable self-use ratio of the community from 72.7% to 98.6%. A similar intercity energy migration framework has been established in THE Guangdong-Hong Kong-Macao Greater Bay Area by Zhou [49]. Both energy pricing policy and energy interaction modes play critical roles in incentivizing the willingness of the stakeholders in accepting the energy migration framework.

5.4.3 Decentralized and centralized battery storages in communities

Decentralized battery storages are always adopted in single buildings with electricity supply for small-scale districts. The techno-economic performance and energy flexibility of decentralized battery storage have been analyzed, and they could be optimized by adjusting battery capacity and electric tariffs [66]. Besides, with the expansion and digitalization of power distribution infrastructures, a peer-to-peer (P2P) energy trading mode in decentralized energy resources has been developed [67], which is able to incentivize the advancement of the applications of renewable energy in communities. Dong et al. [68] presented a P2P energy trading strategy for a decentralized energy blockchain. The strategy was verified by a case study that it could promote the coordination and complementation of decentralized energy storages and improve trading efficiency without sacrificing privacy. Mehdinejad et al. [69] designed a decentralized P2P energy token market for energy trading and developed a new approach called the primal-dual subgradient method for clearing the energy market. The numerical results validated the effectiveness of the strategy.

Centralized battery storages consist of large stationary battery systems that supply electricity for the whole district communities. Compared with decentralized battery storages, large-scale battery systems perform better in grid-related operations in terms of economic feasibility and simulation efficiency [70]. Chauhan et al. [71] comparatively studied the techno-economic performance between centralized and decentralized energy systems in a microgrid. The results indicated that, compared with decentralized energy systems, centralized energy systems can reduce capital costs by 4.71% and energy losses by 1.45%. Furthermore, hybrid energy systems with centralized and decentralized battery storages have also been studied. A predominant coordination and arrangement of decentralized and centralized energy resources in communities can improve techno-economic performance, reduce energy loss, and enhance voltage stability [72].

5.4.4 Building integrated microhydrogen system and hydrogen stations

Nowadays, a lot of research is focused on the integration of microhydrogen systems and hydrogen stations with buildings. The microhydrogen system includes stationary hydrogen storages and FCEVs integrated with buildings. However, the commercial development of the microhydrogen system is limited. On the one hand, because of the high pressure of hydrogen storages, which could be up to hundreds of atmospheres, the security of the energy system has been controversial. On the other hand, sometimes the intermittent renewable energy is hard to surpass the idling power of the electrolyzer, which cannot activate the electrolyzer to operate. Moreover, the energy conversion efficiency of the microhydrogen system is relatively low, such as 60%−70% [73] for electrolyzers and 30%−50% [74] for fuel cells, which highly restricts its applications in supplying electricity from renewable energy.

Hydrogen stations are capable of supplying hydrogen to FCEVs, which adopt the pipeline hydrogen supply and force the pressure to be higher than 700 atmospheres. Recently, some researchers concentrated on the enhancement of hydrogen station security [75]. As shown in Fig. 5.5, hydrogen stations can be supplied by hydrogen pipelines, renewable production by electrolysis, and onsite production from hydrogen feedstock. In traditional methods, there is no interaction between hydrogen stations and buildings, and such limited usage has restrained the development of hydrogen facility applications. He et al. [8] integrated hydrogen stations with micro-energy networks through energy interactions of hydrogen stations-FCEVs-buildings, and developed an energy management strategy to enhance energy flexibility and grid stability. Hydrogen stations could

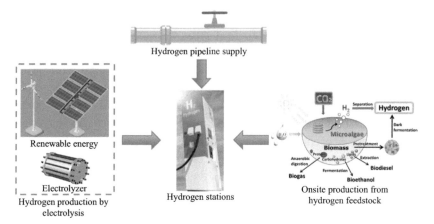

Figure 5.5 Schematic of hydrogen sources of hydrogen stations [78].

reduce public concerns about the security of high-pressure hydrogen storages, and the designed operation could reduce the maximum mean hourly grid power and annual energy cost by 24.2% and 38.9%, respectively. Pang et al. [76] designed an off-grid integrated energy system with a hydrogen station to analyze the optimal size and schedule of the equipment. They verified that PV performance is greatly related to hydrogen generation and storage. Barhoumi et al. [77] compared three configurations of hydrogen stations, including wind-park hydrogen stations with backup batteries, backup fuel cells, and grid connections. The results showed that the wind-park hydrogen station with grid connections was the most cost-efficient.

5.5 Challenges and prospects

5.5.1 Battery and hydrogen storages

Regarding the challenges for battery and hydrogen storages in communities, operation cost, material recycling, and performance durability have been the sustainable research topics, which will fundamentally influence the applications of electrification and hydrogenation in transportations—building systems.

Cost reduction is essential to accelerate commercial development, and some scholars are devoted to developing cost-efficient materials to replace the expensive ones. For instance, cobalt-free high-nickel cathode materials have been designed to reduce the proportion of high-cost cobalt in lithium—ion batteries [79]; copper materials are alloyed with platinum in a fuel cell catalyst to decrease the platinum proportion without sacrificing the activity performance [80]. Considering the environmental protection and reutilization of materials, recycling methods of battery materials have been of concern. Nayaka [81] proposed a hydrometallurgical leaching process for recycling metal ions from lithium—ion battery cathodes, and this method was proven to be efficient and environmentally friendly. Yan et al. [82] analyzed a mechanical crushing method and found that a two-step crushing method could achieve 100% dissociation of all battery components, and improve the efficiency of the recycling process. Jiang et al. [83] quantified the environmental impacts of hydrometallurgical recycling of lithium nickel manganese cobalt oxide and lithium iron phosphate batteries and offered a transparent life cycle inventory for hydrometallurgical recycling of lithium—ion traction batteries for sustainable management.

Performance durability plays a significant role in the techno—economics of community energy networks [50]. Nevertheless, performance degradation indicates different principles in electrification (energy capacity degradation) and hydrogenation (degradation in power conversion efficiency). Regarding the electrification of batteries, a solid electrolyte interface film will be generated on the anode electrode surface. The cathode material structures are deformed and dissolved during charging/discharging cycles [84], leading to energy capacity degradation. Currently, a lot of experiments have been conducted in trying to relieve battery performance degradation [85—89]. Besides, developing the management strategies of battery storage system control has been effective as well, including charge control [90], energy management between different sources [91], restrictions of dynamic operations [92], etc. As for hydrogenation techniques, the energy capacity is dependent on the volumetric and gravimetric density of hydrogen storage, while the output of electric energy is determined by the energy efficiency of fuel cells. The catalyst carbon corrosion, platinum particle aggregation, and membrane crack of fuel cells will lead to the degradation of electric energy [93]. The durability of the fuel cells can be improved by catalyst material redesign [94], operation strategy development [95], energy management [96], etc.

5.5.2 Decentralized and centralized energy storage systems

The management of a decentralized and centralized energy storage system, energy flexibility, and energy efficiency are formidable challenges.

The decentralized energy system in the community can be controlled easily and flexibly for optimal energy management. Han et al. [97] proposed a decentralized energy management strategy for an islanded PV/hydrogen/battery system. The strategy was verified in a simulated platform that had an efficient power distribution system. However, because of the intermittent renewable energy generation and random electricity load variation, there will be frequent charging/discharging cycles of batteries, which has a negative effect on battery lifetime [62]. On the contrary, the centralized energy system possesses less charging/discharging cycles because of the offset between the aggregated renewable power and the accumulated electricity load. The control strategy of the centralized energy system is less flexible compared with that ofthe decentralized energy system because different requirements and willingness from different building owners need to be considered and balanced.

Fig. 5.6 shows the energy route and rough energy efficiency of the battery storage system. The energy efficiency of traditional energy resources converted to electric energy, including coal, oil, and natural gas, is usually higher than that of renewable energy resources, such as solar energy. While the maximum efficiency of charging/discharging for battery storages is around 95%–98% [14], the overall energy efficiency of traditional energy sources is about 36.1%, higher than that of renewable energy efficiency of about 18.1%.

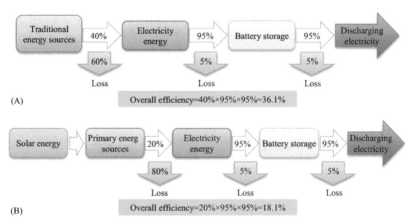

Figure 5.6 The energy route and rough energy efficiency of battery storage systems. (A) Traditional energy (coal, oil, and natural gas) for primary energy sources; (B) solar energy for primary energy sources.

5.5.3 Microhydrogen system and hydrogen stations

The advantages of nonpollution and zero carbon emissions of micro-hydrogen systems and hydrogen stations result in great potential for their widespread applications in buildings and transportations. However, the security of hydrogen usage and energy efficiency needs to be improved.

Concerns with microhydrogen integrated building energy systems include high-pressure hydrogen leakage, conflagration, and explosion [98]. The liquid hydrogen storage technology has been generalized [99−101] for volumetric power density improvement, leading to a significant increase in energy security studies. Temperature control is one of the promising methods, and it can compensate for the thermal kinetic energy, and relieve the risk of hydrogen leakage [99]. The behavior of hydrogen leakage and diffusion has a critical role in microhydrogen system security. Qian et al. [102] compared six scenarios of different hydrogen leakage positions and wind effects, based on which the space arrangement of the hydrogen storage system can be optimized. As for the hydrogen stations, materials of critical components play an essential role. Due to the hydrogen embrittlement effect, the ductility, fracture resistance, and fatigue properties of materials touching high-pressure hydrogen will be adversely changed [13]. Thus the hydrogen compatibility of materials is a formidable challenge. Hao et al. [103] conducted experiments to analyze the hydrogen compatibility of different rubber seals used in hydrogen stations in a 700-atmosphere hydrogen environment and provided guidance for material choice.

At the system level, the characteristic of low system energy efficiency blocks commercial development. Fig. 5.7 illustrates the energy route and rough energy efficiency of hydrogen systems. Due to the low efficiency of hydrogen transferring to electricity, which is about 30%−50% [74], the overall electricity conversion efficiency of hydrogen systems is around 11.2% for traditional energy resources (coal, oil, and natural gas), and it is much lower compared with that of battery storage systems. When the primary energy source is solar energy, the overall electricity conversion efficiency is only around 5.6%. Thus the PEMFC CHP cogeneration system has huge potential in building applications as it takes advantage of thermal energy for supplying hot water, and improves hydrogen conversion efficiency by more than 85% [104], with system efficiency reaching about 11.9% overall.

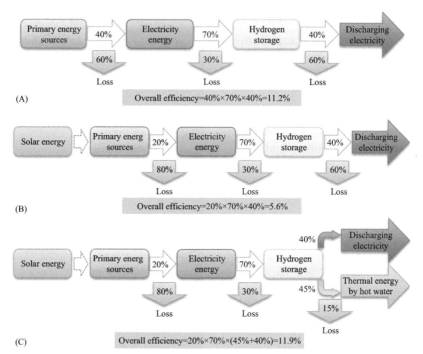

Figure 5.7 The energy route and rough energy efficiency of hydrogen systems. (A) Traditional energy (coal, oil, and natural gas) for primary energy sources; (B) solar energy for primary energy sources; (C) Hydrogen CHP cogeneration systems supported by solar energy. *CHP*, Combined heat and power.

5.5.4 Potential frameworks and prospects of low-carbon communities with electrification and hydrogenation

Two potential frameworks of low-carbon energy communities with electrification and hydrogenation, that is, building-scale-based district energy sharing networks and an intercity energy migration framework, are discussed in this section.

Fig. 5.8 demonstrates the building-scale based district energy sharing network. Building-A is capable of producing electricity through building integrated photovoltaics (BIPVs) to fulfill its power demand and also supply power for the charging of EVs and hydrogen production by the electrolyzer. The hydrogen tank can store the produced hydrogen from the electrolyzer and pipeline hydrogen from the gas network, and supply it to FCEVs and fuel cells for CHP cogeneration, which can support the energy demand in building B. Building-scale-based district energy-sharing

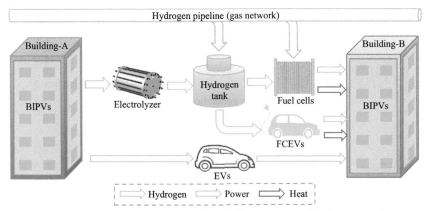

Figure 5.8 Schematic of building scale-based district energy-sharing networks.

networks can improve the ratio of renewable penetration and decarbonization in communities, and the independence of energy self-sufficiency of communities will also be enhanced. Besides, such frameworks make it possible to conduct P2P energy-sharing trading in communities.

Fig. 5.9 demonstrates the intercity energy migration framework. Remote regions, such as suburban areas, possess great potential for renewable energy generation, but have low energy consumption demands. The BIPVs, rooftop PVs, and wind turbines produce electric energy for the charging of EVs and hydrogen generation by the electrolyzer [105], as well as support for FCEV charging by the hydrogen station. EVs and FCEVs transport renewable energy to the city center by daily transportations, where there is a great demand for energy consumption. In the city center, the energy interaction between the EVs, FCEVs, buildings, and grid can be controlled for optimal techno-economics. The intercity energy migration framework benefits the energy distribution balance between regions with different energy generation and demand and improves the efficiency of the energy system and the decarbonization of city communities [106,107].

A number of technical challenges still exist to achieve electrification and hydrogenation in the building-transportation system in a matured way. It is necessary to put efforts to the following research aspects: (1) reduce the cost of industrial large-scale hydrogen generation and production of electrochemical batteries and fuel cells through the development and implementation of high-efficiency but low-cost material; (2) improve the durability and life cycle of electrochemical batteries and fuel cells through the development of high-durability material and system control strategy; (3) improve energy flexibility for community renewable energy networks by advanced strategy

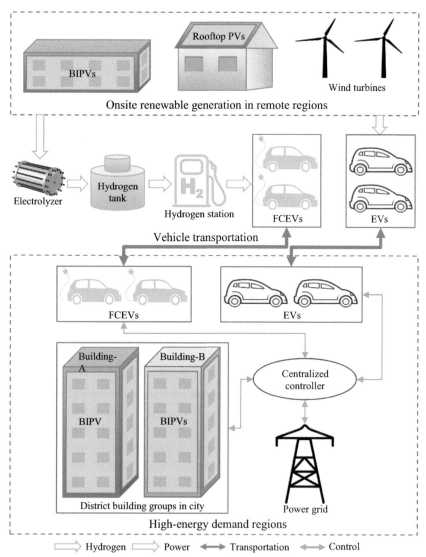

Figure 5.9 Schematic of the intercity energy migration framework.

control; (4) optimize the system energy efficiency of battery and hydrogen energy systems; (5) improve energy system security by developing suitable hydrogen–compatibility materials and designing a proper energy system; (6) develop economic incentive policies for raising the acceptance of energy sharing and migration frameworks to stakeholders; (7) advanced machine learning in sustainable and smart cities [108,109].

Acknowledgment

This work was supported by Regional joint fund youth fund project (2022A1515110364, P00038-1002), Guangdong Basic and Applied Basic Research Foundation 2023 (2023A04J1035, P00121-1003), Joint Funding of Institutes and Enterprises in 2023 (2023A03J0104, P00054-1003,1004), HKUST(GZ)-enterprise cooperation project (R00017-2001), and HKUST(GZ)-enterprise cooperation project "Optimization Design of Proton Exchange Membrane Fuel Cell Plate" (R00072-2001). This research is supported by The Hong Kong University of Science and Technology (Guangzhou) startup grant (G0101000059). This work was also supported in part by the Project of Hetao Shenzhen-Hong Kong Science and Technology Innovation Cooperation Zone (HZQB-KCZYB-2020083).

References

[1] Masiero G, Ogasavara MH, Jussani AC, Risso ML. Electric vehicles in China: BYD strategies and government subsidies. RAI Revista de Administração e Inovação 2016;13:3−11.

[2] Yoshida T, Kojima K. Toyota MIRAI fuel cell vehicle and progress toward a future hydrogen society. The Electrochemical Society Interface 2015;24:45.

[3] Giménez SN, Durá JMH, Ferragud FXB, Fernández RS. Control-oriented modeling of the cooling process of a PEMFC-based μ-CHP system. IEEE Access 2019;7:95620−42.

[4] Mehrtash M, Capitanescu F, Heiselberg PK, Gibon T, Bertrand A. An enhanced optimal pv and battery sizing model for zero energy buildings considering environmental impacts. IEEE Transactions on Industry Applications 2020;56:6846−56.

[5] Rosati A, Facci AL, Ubertini S. Techno-economic analysis of battery electricity storage towards self-sufficient buildings. Energy Conversion and Management 2022;256:115313.

[6] Pandžić H. Optimal battery energy storage investment in buildings. Energy and Buildings 2018;175:189−98.

[7] Wang D, Muratori M, Eichman J, Wei M, Saxena S, Zhang C. Quantifying the flexibility of hydrogen production systems to support large-scale renewable energy integration. Journal of Power Sources 2018;399:383−91.

[8] He Y, Zhou Y, Yuan J, Liu Z, Wang Z, Zhang G. Transformation towards a carbon-neutral residential community with hydrogen economy and advanced energy management strategies. Energy Conversion and Management 2021;249.

[9] Qian F, Gao W, Yang Y, Yu D. Economic optimization and potential analysis of fuel cell vehicle-to-grid (FCV2G) system with large-scale buildings. Energy Conversion and Management 2020;205.

[10] Farzin H, Fotuhi-Firuzabad M, Moeini-Aghtaie M. A practical scheme to involve degradation cost of lithium-ion batteries in vehicle-to-grid applications. IEEE Transactions on Sustainable Energy 2016;7:1730−8.

[11] Thompson AW. Economic implications of lithium ion battery degradation for Vehicle-to-Grid (V2X) services. Journal of Power Sources 2018;396:691−709.

[12] Uddin K, Jackson T, Widanage WD, Chouchelamane G, Jennings PA, Marco J. On the possibility of extending the lifetime of lithium-ion batteries through optimal V2G facilitated by an integrated vehicle and smart-grid system. Energy. 2017;133:710−22.

[13] Wang H, Tong Z, Zhou G, Zhang C, Zhou H, Wang Y, et al. Research and demonstration on hydrogen compatibility of pipelines: a review of current status and challenges. International Journal of Hydrogen Energy 2022;47:28585−604.

[14] Wang J, Wang B, Zhang L, Wang J, Shchurov NI, Malozyomov BV. Review of bidirectional DC−DC converter topologies for hybrid energy storage system of new energy vehicles. Green Energy and Intelligent Transportation 2022;1.

[15] Kumar D, Rajouria SK, Kuhar SB, Kanchan DK. Progress and prospects of sodium-sulfur batteries: a review. Solid State Ionics 2017;312:8−16.

[16] Pelletier S, Jabali O, Laporte G, Veneroni M. Battery degradation and behaviour for electric vehicles: review and numerical analyses of several models. Transportation Research Part B Methodological, 2017;103:158−87.

[17] Singh A, Karandikar PB, Kulkarni NR. Mitigation of sulfation in lead acid battery towards life time extension using ultra capacitor in hybrid electric vehicle. Journal of Energy Storage 2021;34:102219.

[18] Manzetti S, Mariasiu F. Electric vehicle battery technologies: from present state to future systems. Renewable and Sustainable Energy Reviews 2015;51:1004−12.

[19] Zhou Y, Cao S, Hensen JLM, Lund PD. Energy integration and interaction between buildings and vehicles: a state-of-the-art review. Renewable and Sustainable Energy Reviews 2019;114.

[20] Hu X, Zou C, Zhang C, Li Y. Technological developments in batteries: a survey of principal roles, types, and management needs. IEEE Power and Energy Magazine 2017;15:20−31.

[21] Zhou Y. Transition towards carbon-neutral districts based on storage techniques and spatiotemporal energy sharing with electrification and hydrogenation. Renewable and Sustainable Energy Reviews 2022;162.

[22] Bullock KR. Lead/acid batteries. Journal of Power Sources 1994;51:1−17.

[23] Fetcenko MA, Ovshinsky SR, Reichman B, Young K, Fierro C, Koch J, et al. Recent advances in NiMH battery technology. Journal of Power Sources 2007;165:544−51.

[24] Petrovic S. Nickel−cadmium batteries. Battery Technology Crash Course 2021;73−88.

[25] Accardo A, Dotelli G, Musa ML, Spessa EJAS. Life cycle assessment of an NMC battery for application to electric light-duty commercial vehicles and comparison with a sodium-nickel-chloride battery. Applied Sciences 2021;11:1160.

[26] Bracco S, Delfino F, Trucco A, Zin S. Electrical storage systems based on Sodium/Nickel chloride batteries: a mathematical model for the cell electrical parameter evaluation validated on a real smart microgrid application. Journal of Power Sources 2018;399:372-82.

[27] Capasso C, Lauria D, Veneri O. Experimental evaluation of model-based control strategies of sodium-nickel chloride battery plus supercapacitor hybrid storage systems for urban electric vehicles. Applied Energy 2018;228:2478-89.

[28] Longo S, Cellura M, Cusenza MA, Guarino F, Mistretta M, Panno D, et al. Life cycle assessment for supporting eco-design: The case study of sodium−nickel chloride cells. Energies 2021;14:1897.

[29] Sudworth JL. The sodium/nickel chloride (ZEBRA) battery. Journal of Power Sources 2001;100:149−63.

[30] Liu C, Neale ZG, Cao G. Understanding electrochemical potentials of cathode materials in rechargeable batteries. Materials Today 2016;19:109−23.

[31] Sepúlveda-Mora SB, Hegedus S. Making the case for time-of-use electric rates to boost the value of battery storage in commercial buildings with grid connected PV systems. Energy. 2021;218:119447.

[32] Alqaed S. Heating a residential building using the heat generated in the lithium ion battery pack by the electrochemical process. Journal of Energy Storage 2022;45:103553.

[33] Shiva Kumar S, Lim H. An overview of water electrolysis technologies for green hydrogen production. Energy Reports 2022;8:13793−813.

[34] Catumba BD, Sales MB, Borges PT, Ribeiro Filho MN, Lopes AAS, Sousa Rios MAd, et al. Sustainability and challenges in hydrogen production: an advanced bibliometric analysis. International Journal of Hydrogen Energy 2022.

[35] Schmidt O, Gambhir A, Staffell I, Hawkes A, Nelson J, Few S. Future cost and performance of water electrolysis: an expert elicitation study. International Journal of Hydrogen Energy 2017;42:30470−92.

[36] Kang SY, Park JE, Jang GY, Kim O-H, Kwon OJ, Cho Y-H, et al. High-performance and durable water electrolysis using a highly conductive and stable anion-exchange membrane. International Journal of Hydrogen Energy 2022;47:9115−26.

[37] Santos DM, Sequeira CA, Figueiredo J. Hydrogen production by alkaline water electrolysis. Química Nova 2013;36:1176−93.

[38] Wang H, Zhao Y, Dong X, Yang J, Guo H, Gong M. Thermodynamic analysis of low-temperature and high-pressure (cryo-compressed) hydrogen storage processes cooled by mixed-refrigerants. International Journal of Hydrogen Energy 2022;47:28932−44.

[39] Yan Y, Xu Z, Han F, Wang Z, Ni Z. Energy control of providing cryo-compressed hydrogen for the heavy-duty trucks driving. Energy. 2022;242:122817.

[40] Xu Z, Yan Y, Wei W, Sun D, Ni Z. Supply system of cryo-compressed hydrogen for fuel cell stacks on heavy duty trucks. International Journal of Hydrogen Energy 2020;45:12921−31.

[41] Yanxing Z, Maoqiong G, Yuan Z, Xueqiang D, Jun S. Thermodynamics analysis of hydrogen storage based on compressed gaseous hydrogen, liquid hydrogen and cryo-compressed hydrogen. International Journal of Hydrogen Energy 2019;44:16833−40.

[42] Larpruenrudee P, Bennett NS, Gu Y, Fitch R, Islam MS. Design optimization of a magnesium-based metal hydride hydrogen energy storage system. Scientific Reports 2022;12:13436.

[43] Sathe RY, Dhilip Kumar TJ, Ahuja R. Furtherance of the material-based hydrogen storage based on theory and experiments. International Journal of Hydrogen Energy 2023.

[44] Xu Y, Deng Y, Liu W, Zhao X, Xu J, Yuan Z. Research progress of hydrogen energy and metal hydrogen storage materials. Sustainable Energy Technologies and Assessments 2023;55:102974.

[45] Sui Y, Yuan Z, Zhou D, Zhai T, Li X, Feng D, et al. Recent progress of nanotechnology in enhancing hydrogen storage performance of magnesium-based materials: a review. International Journal of Hydrogen Energy 2022;47:30546−66.

[46] Samimi F, Ghiyasiyan-Arani M, Salavati-Niasari M, Alwash SW. A study of relative electrochemical hydrogen storage capacity of active materials based on $Zn_3Mo_2O_9$/ZnO and $Zn_3Mo_2O_9$/$ZnMoO_4$. International Journal of Hydrogen Energy 2022.

[47] Yang Q, Gao B, Xiao G, Hao D, Chen S. Transient model of dynamic power output under PEMFC load current variations. International Journal of Green Energy 2022;19:1543−53.

[48] Ou K, Yuan W-W, Kim Y-B. Development of optimal energy management for a residential fuel cell hybrid power system with heat recovery. Energy. 2021;219:119499.

[49] Zhou Y. Energy sharing and trading on a novel spatiotemporal energy network in Guangdong-Hong Kong-Macao Greater Bay Area. Applied Energy 2022;318.

[50] He Y, Zhou Y, Wang Z, Liu J, Liu Z, Zhang G. Quantification on fuel cell degradation and techno-economic analysis of a hydrogen-based grid-interactive residential energy sharing network with fuel-cell-powered vehicles. Applied Energy 2021;303.

[51] He Y, Zhou Y, Liu J, Liu Z, Zhang G. An inter-city energy migration framework for regional energy balance through daily commuting fuel-cell vehicles. Applied Energy 2022;324.

[52] Li Y, Gao W, Ruan Y. Performance investigation of grid-connected residential PV-battery system focusing on enhancing self-consumption and peak shaving in Kyushu, Japan. Renewable Energy 2018;127:514−23.

[53] Mahmud K, Hossain MJ, Town GE. Peak-load reduction by coordinated response of photovoltaics, battery storage, and electric vehicles. IEEE Access 2018;6:29353−65.

[54] Kottick D, Blau M, Edelstein D. Battery energy storage for frequency regulation in an island power system. IEEE Transactions on Energy Conversion 1993;8:455−9.

[55] He G, Chen Q, Kang C, Xia Q, Poolla K. Cooperation of wind power and battery storage to provide frequency regulation in power markets. IEEE Transactions on Power Systems 2017;32:3559−68.

[56] Shi Y, Xu B, Wang D, Zhang B. Using battery storage for peak shaving and frequency regulation: joint optimization for superlinear gains. IEEE Transactions on Power Systems 2018;33:2882−94.

[57] Peng C, Zou J, Lian L. Dispatching strategies of electric vehicles participating in frequency regulation on power grid: a review. Renewable and Sustainable Energy Reviews 2017;68:147−52.

[58] Han X, Ji T, Zhao Z, Zhang H. Economic evaluation of batteries planning in energy storage power stations for load shifting. Renewable Energy 2015;78:643−7.

[59] Tang H, Wang S. Life-cycle economic analysis of thermal energy storage, new and second-life batteries in buildings for providing multiple flexibility services in electricity markets. Energy. 2023;264:126270.

[60] Taie Z, Villaverde G, Speaks Morris J, Lavrich Z, Chittum A, White K, et al. Hydrogen for heat: using underground hydrogen storage for seasonal energy shifting in northern climates. International Journal of Hydrogen Energy 2021;46:3365−78.

[61] Kilic M, Altun AF. Dynamic modelling and multi-objective optimization of off-grid hybrid energy systems by using battery or hydrogen storage for different climates. International Journal of Hydrogen Energy 2022.

[62] Zhou Y, Cao S, Hensen JLM, Hasan A. Heuristic battery-protective strategy for energy management of an interactive renewables−buildings−vehicles energy sharing network with high energy flexibility. Energy Conversion and Management 2020;214:112891.

[63] Zhou Y, Cao S, Kosonen R, Hamdy M. Multi-objective optimisation of an interactive buildings-vehicles energy sharing network with high energy flexibility using the Pareto archive NSGA-II algorithm. Energy Conversion and Management 2020;218:113017.

[64] Zhou Y, Cao S, Hensen JLM. An energy paradigm transition framework from negative towards positive district energy sharing networks—battery cycling aging, advanced battery management strategies, flexible vehicles-to-buildings interactions, uncertainty and sensitivity analysis. Applied Energy 2021;288:116606.

[65] Robledo CB, Oldenbroek V, Abbruzzese F, van Wijk AJM. Integrating a hydrogen fuel cell electric vehicle with vehicle-to-grid technology, photovoltaic power and a residential building. Applied Energy 2018;215:615−29.

[66] Zhou Y, Cao S. Energy flexibility investigation of advanced grid-responsive energy control strategies with the static battery and electric vehicles: a case study of a high-rise office building in Hong Kong. Energy Conversion and Management 2019;199:111888.

[67] Umar A, Kumar D, Ghose T. Blockchain-based decentralized energy intra-trading with battery storage flexibility in a community microgrid system. Applied Energy 2022;322:119544.

[68] Dong J, Song C, Liu S, Yin H, Zheng H, Li Y. Decentralized peer-to-peer energy trading strategy in energy blockchain environment: a game-theoretic approach. Applied Energy 2022;325:119852.

[69] Mehdinejad M, Shayanfar H, Mohammadi-Ivatloo B. Decentralized blockchain-based peer-to-peer energy-backed token trading for active prosumers. Energy. 2022;244:122713.

[70] Zeh A, Rau M, Witzmann R. Comparison of decentralised and centralised grid-compatible battery storage systems in distribution grids with high PV penetration. Progress in Photovoltaics: Research and Applications 2016;24:496−506.

[71] Chauhan RK, Chauhan K, Subrahmanyam BR, Singh AG, Garg MM. Distributed and centralized autonomous DC microgrid for residential buildings: a case study. Journal of Building Engineering 2020;27:100978.

[72] Ahmadi M, Adewuyi OB, Danish MSS, Mandal P, Yona A, Senjyu T. Optimum coordination of centralized and distributed renewable power generation incorporating battery storage system into the electric distribution network. International Journal of Electrical Power & Energy Systems 2021;125:106458.

[73] Zhang H, Su S, Lin G, Chen J. Efficiency calculation and configuration design of a PEM electrolyzer system for hydrogen production. International Journal of Electrochemical Science 2012;7:4143−57.

[74] Mekhilef S, Saidur R, Safari A. Comparative study of different fuel cell technologies. Renewable and Sustainable Energy Reviews 2012;16:981−9.

[75] Kwon D, Choi SK, Yu C. Improved safety by crossanalyzing quantitative risk assessment of hydrogen refueling stations. International Journal of Hydrogen Energy 2022;47:10788−98.

[76] Pang Y, Pan L, Zhang J, Chen J, Dong Y, Sun H. Integrated sizing and scheduling of an off-grid integrated energy system for an isolated renewable energy hydrogen refueling station. Applied Energy 2022;323:119573.

[77] Barhoumi EM, Salhi MS, Okonkwo PC, Ben Belgacem I, Farhani S, Zghaibeh M, et al. Techno-economic optimization of wind energy based hydrogen refueling station case study Salalah city Oman. International Journal of Hydrogen Energy 2022.

[78] Wang J, Yin Y. Fermentative hydrogen production using pretreated microalgal biomass as feedstock. Microbial Cell Factories 2018;17:22.

[79] Noerochim L, Suwarno S, Idris NH, Dipojono HK. Recent development of nickel-rich and cobalt-free cathode materials for lithium-ion batteries. Batteries 2021.

[80] Yılmaz MS, Kaplan BY, Gürsel SA, Metin Ö. Binary CuPt alloy nanoparticles assembled on reduced graphene oxide-carbon black hybrid as efficient and cost-effective electrocatalyst for PEMFC. International Journal of Hydrogen Energy 2019;44:14184−92.

[81] Nayaka GP, Zhang Y, Dong P, Wang D, Zhou Z, Duan J, et al. An environmental friendly attempt to recycle the spent Li-ion battery cathode through organic acid leaching. Journal of Environmental Chemical Engineering 2019;7:102854.

[82] Yan B, Ma E, Wang J. Research on the high-efficiency crushing, sorting and recycling process of column-shaped waste lithium batteries. Science of the Total Environment 2023;864:161081.

[83] Jiang S, Hua H, Zhang L, Liu X, Wu H, Yuan Z. Environmental impacts of hydro-metallurgical recycling and reusing for manufacturing of lithium-ion traction batteries in China. Science of the Total Environment 2022;811:152224.

[84] Han X, Lu L, Zheng Y, Feng X, Li Z, Li J, et al. A review on the key issues of the lithium ion battery degradation among the whole life cycle. eTransportation 2019;1.
[85] Romano Brandt L, Nishio K, Salvati E, Simon KP, Papadaki C, Hitosugi T, et al. Improving ultra-fast charging performance and durability of all solid state thin film Li-NMC battery-on-chip systems by in situ TEM lamella analysis. Applied Materials Today 2022;26:101282.
[86] Lamberti A, Garino N, Sacco A, Bianco S, Manfredi D, Gerbaldi C. Vertically aligned TiO2 nanotube array for high rate Li-based micro-battery anodes with improved durability. Electrochimica Acta 2013;102:233−9.
[87] Chen Y, He T, Liu Q, Hu Y, Gu H, Deng L, et al. Highly durable iron single-atom catalysts for low-temperature zinc-air batteries by electronic regulation of adjacent iron nanoclusters. Applied Catalysis B: Environmental 2023;323:122163.
[88] Zhang M, Hu X, Xin Y, Wang L, Zhou Z, Yang L, et al. FeNi coordination polymer based highly efficient and durable bifunction oxygen electrocatalyst for rechargeable zinc-air battery. Separation and Purification Technology 2023;308:122974.
[89] Nagaraju G, Santhoshkumar P, Sekhar SC, Ramulu B, Nanthagopal M, Babu PSS, et al. Metal organic framework-derived MnO@carbon composites for highly durable Li-ion batteries and hybrid electrochemical cells. Journal of Power Sources 2022;549:232113.
[90] Du J, Liu Y, Mo X, Li Y, Li J, Wu X, et al. Impact of high-power charging on the durability and safety of lithium batteries used in long-range battery electric vehicles. Applied Energy 2019;255:113793.
[91] Hu J, Wang Z, Du H, Zou L. Hierarchical energy management strategy for fuel cell/ ultracapacitor/battery hybrid vehicle with life balance control. Energy Conversion and Management 2022;272:116383.
[92] Desantes JM, Novella R, Pla B, Lopez-Juarez M. Effect of dynamic and operational restrictions in the energy management strategy on fuel cell range extender electric vehicle performance and durability in driving conditions. Energy Conversion and Management 2022;266:115821.
[93] Jahnke T, Futter G, Latz A, Malkow T, Papakonstantinou G, Tsotridis G, et al. Performance and degradation of proton exchange membrane fuel cells: state of the art in modeling from atomistic to system scale. Journal of Power Sources 2016;304:207−33.
[94] Prithi JA, Shanmugam R, Sahoo MK, Rajalakshmi N, Rao GR. Evaluation of the durability of ZrC as support material for Pt electrocatalysts in PEMFCs: experimental and computational studies. International Journal of Hydrogen Energy 2022;47:36232−47.
[95] Yang Q, Gao B, Cheng Q, Xiao G, Meng M. Adaptive control strategy for power output stability in long-time operation of fuel cells. Energy. 2022;238.
[96] Sheng C, Fu J, Li D, Jiang C, Guo Z, Li B, et al. Energy management strategy based on health state for a PEMFC/Lithium-ion batteries hybrid power system. Energy Conversion and Management 2022;271:116330.
[97] Han Y, Yang H, Li Q, Chen W, Zare F, Guerrero JM. Mode-triggered droop method for the decentralized energy management of an islanded hybrid PV/hydrogen/battery DC microgrid. Energy. 2020;199:117441.
[98] Hord J. Is hydrogen a safe fuel? International Journal of Hydrogen Energy 1978;3:157−76.
[99] Wan C, Zhu S, Shi C, Bao S, Zhi X, Qiu L, et al. Numerical simulation on pressure evolution process of liquid hydrogen storage tank with active cryogenic cooling: simulation numérique du processus d'évolution de pression du réservoir de stockage d'hydrogène liquide avec refroidissement cryogénique actif. International Journal of Refrigeration 2023.

[100] Lee H, Ryu B, Anh DP, Roh G, Lee S, Kang H. Thermodynamic analysis and assessment of novel ORC-DEC integrated PEMFC system for liquid hydrogen fueled ship application. International Journal of Hydrogen Energy 2023;48:3135−53.

[101] Lee J, Park B, Kim K-H, Ruy W-S. Multi-objective optimization of liquid hydrogen FPSO at the conceptual design stage. International Journal of Naval Architecture and Ocean Engineering 2023;15:100511.

[102] Qian J-y, Li X-j, Gao Z-x, Jin Z-j. A numerical study of hydrogen leakage and diffusion in a hydrogen refueling station. International Journal of Hydrogen Energy 2020;45:14428−39.

[103] Hao D, Yang Y, Gao B, Zhang S, Zhang Y, Wang R. An experimental study on high-pressure hydrogen compatibility of rubber seals at FCV refuelling receptacles. IOP Conference Series: Earth and Environmental Science. 2021;714.

[104] Zou W-J, Shen K-Y, Jung S, Kim Y-B. Application of thermoelectric devices in performance optimization of a domestic PEMFC-based CHP system. Energy. 2021;229:120698.

[105] Zhou L, Zhou Y. Study on thermo-electric-hydrogen conversion mechanisms and synergistic operation on hydrogen fuel cell and electrochemical battery in energy flexible buildings. Energy Conversion and Management 2023;116610. In this issue.

[106] Zhou Y. Sustainable energy sharing districts with electrochemical battery degradation in design, planning, operation and multi-objective optimisation. Renewable Energy 2023;202:1324−41. In this issue.

[107] Zhou Y. Climate change adaptation with energy resilience in energy districts—A state-of-the-art review. Energy and Buildings 2023;112649.

[108] Zhou Y. Advances of machine learning in multi-energy district communities-mechanisms, applications and perspectives. Energy and AI 2022;100187. In this issue.

[109] Zhou Y. Artificial intelligence in renewable systems for transformation towards intelligent buildings. Energy and AI 2022;100182. In press.

CHAPTER 6

Hybrid energy storages in buildings with artificial intelligence

Ying Sun[1] and Zhengxuan Liu[2]
[1]School of Environmental and Municipal Engineering, Qingdao University of Technology, Qingdao, P.R. China
[2]Faculty of Architecture and the Built Environment, Delft University of Technology, Delft, The Netherlands

6.1 Introduction

Due to the ability to store energy for later usage, energy storage systems (ESSs) have drawn massive attention in building engineering for peak shifting or peak shaving in recent years [1,2]. The International Energy Agency reported that the building and building construction sector accounted for almost one third of the global final energy consumption in the year 2021 [3]. The energy consumed by buildings tends to continually increase in the following years. Besides, occupants' living habits may result in a relatively higher energy demand during certain periods. For instance, the cooking and heating demand of occupants during winter may create daily peaks in energy consumption during early mornings and late afternoons [4]. The intensive energy demand during peak periods would cause increased stress on the grid for electricity distribution. Therefore using ESSs to store energy during off-peak periods and release it to the consumer during peak periods would be helpful in easing the burden on the main grids, as the electricity imported from the grids could be effectively decreased [5,6,7].

On the other hand, ESS in buildings could be beneficial for solving the intermittent problem of renewable energy sources and achieving self-sufficiency [8]. In recent years, the goal of sustainable development and carbon neutrality has drawn intense research interest in renewable energy generation [9,10,11]. However, renewable energy sources, such as solar and wind energy, are usually intermittent and difficult to be continually utilized [12,13]. Besides, the intermittent availability of renewable sources may result in a mismatch between energy supply and consumption [14].

Advances in Digitalization and Machine Learning for Integrated Building-Transportation Energy Systems
DOI: https://doi.org/10.1016/B978-0-443-13177-6.00004-7

By taking advantage of the energy storage capacity of ESS, the excess renewable energy could be stored and supplied to consumers when needed. In other words, ESS could be an efficient solution in extending the renewable energy share, and thereby achieving energy self-sufficiency [15].

Nowadays various energy storage techniques are available, which could be classified into five categories [16−19]: mechanical, electrochemical, electrical, thermal, and chemical. Plenty of studies have been done to integrate ESS into buildings. Studies on ESSs are usually aimed at optimizing renewable energy storage and battery operation [20]. For instance, Rathore et al. [21] reviewed studies on integrating phase change material (PCM) into building thermal envelopes to form thermal energy storage (TES) systems. Olsthoorn et al. [22] reviewed the modeling and optimization objectives of district heating systems integrated with renewable energy storage technologies. The possibilities of utilizing thermal mass activation in buildings to achieve peak shifting or peak shaving have also been reviewed [23]. Furthermore, Thieblemont et al. [24] summarized supervisory predictive control strategies to properly control the energy storage process in buildings based on weather forecasts.

With the rapid development of smart sensors and the Internet of Things (IoT) [25], plenty of data could be collected to equip buildings with data-rich ESS systems [26]. The huge volume of available datasets encouraged the application of artificial intelligence (AI) in predicting, optimizing, and controlling buildings [27] and ESS systems [28]. In recent years, the applications of AI to predict building energy consumption have been comprehensively reviewed [29−33]. Gao and Lu [34] reviewed the commonly used machine learning (ML) algorithms, including unsupervised ML, supervised ML, and RL (reinforcement learning), for energy storage devices, such as battery and fuel cells, and systems, such as TES systems, microgrids that contain energy storage units, and electric vehicles (EVs). They summarized the existing studies on the application of ML to energy storage devices in terms of status prediction, fault and defect diagnosis, and design and optimization, while the studies on applying ML to ESSs to optimize the control strategy have also been reviewed.

Most AI-based prediction models are supervised ML models, in which the training data are labeled. Commonly used AI-based prediction models include decision tree (DT), artificial neural network (ANN), support vector machine (SVM), deep learning (DL), ensemble method, etc. Besides, AI-based optimization algorithms refer to optimization methods inspired by human brains or animal behaviors. AI-based optimization methods include genetic algorithm (GA), particle swarm optimization (PSO), differential evolution (DE), ant colony optimization (ACO), simulated annealing (SA), immune algorithm (IA), artificial fish swarm algorithm (AFSA), etc. Moreover, the AI-based control strategies could be classified as predictive-based control and predictive-free control.

This chapter introduces commonly used ESS and AI technologies in buildings in Section 6.2. The applications of AI to ESS in buildings are reviewed in Section 6.3. Section 6.4 presents the discussion of the drawbacks of existing case studies and identifies potential future works. Finally, the conclusion is given in Section 6.5.

6.2 Commonly used energy storage systems and artificial intelligence technologies in buildings

6.2.1 Energy storage systems

ESSs, which store energy for later use, not only address the intermittency problem of renewable sources but also enhance the grid stability and resilience by solving the mismatch between energy supply and demand [35,36]. Generally, ESS could be classified into five types based on the energy storage medium [16−19]: mechanical, electrochemical, electrical, thermal, and chemical. A detailed classification is shown in Fig. 6.1. Although various ESSs are available, this section only summarizes commonly studied storage systems in buildings, such as water tanks, batteries, building-integrated TES (BITES) systems, and underground/underwater ESSs.

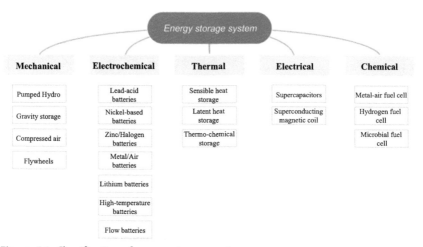

Figure 6.1 Classification of energy storage systems.

6.2.1.1 Water tank

A water tank can be used as a solar storage system [37], which is usually connected to a solar collector to produce and store warm water for space heating and tap water. Large water tanks and gravel-water pits could be utilized to achieve seasonal energy storage that collects energy during the summer and releases it during the winter [38,39].

Water tank systems could also be used to store energy from renewable sources or waste heat. Li et al. [40] integrated a water tank TES system into a heat prosumer to shave the peak load by ~39% by increasing the waste heat self-utilization rate by ~17%. Lyu et al. [41] used a water tank to store heat produced by an air-source heat pump. Their study found that the water tank could effectively reduce the start-stop loss.

On the other hand, water tanks can be used to store cooling power. Zeng et al. [42] proposed a buried water-PCM tank system (see Fig. 6.2) to act as a TES system that provides cooling during peak periods. By taking advantage of the thermal inertia of the surrounding soil, the simulation result shows that the cooling capacity of the proposed buried water-PCM tank is ~25% higher than an insulated tank. Later, to improve the flexibility of the proposed tank and make it easy for mass production, Zeng et al. [43] proposed the concept of multimodular water-PCM tanks (MMWPT) to connect multiple water tanks in a matrix form. The MMWPT was then

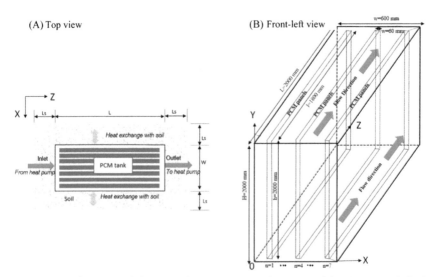

Figure 6.2 Schematic of the buried water-PCM tank proposed by Zeng et al. [42]. (A) Top view. (B) Front-left view. *PCM*, Phase change material.

integrated into a geothermal heat pump system to provide additional cooling under emergency mode [44]. McKenna et al. [45] utilized the latent energy storage capacity of water in a tank to store cooling from a ground source heat pump and release it to cool a commercial building by air handling units.

6.2.1.2 Battery

In recent years, battery ESSs have been a hot topic in building engineering because of their ability to provide the backup for intermittent renewable sources and contribute to energy management. Commonly used battery technologies include lithium–ion, lead–acid, sodium–sulfur, flow, and supercapacitor batteries [46].

Lithium–ion batteries have been widely used in storing electricity generated by solar photovoltaics (PVs). To improve the thermal management performance of lithium–ion batteries in terms of achieving a lower battery surface temperature, Leng et al. [47] investigated HP/PCM-coupled thermal management. Then, they went on to optimize an improved HP/PCM-coupled thermal management that uses fins to enhance the air condensing effect of the HP [48]. The optimization objective is to minimize the PCM thickness and fan power for the worst working conditions.

To make an economic analysis for buildings with solar PVs and battery ESSs, Akter et al. [49] proposed a comprehensive framework that uses different economic measures for economic evaluation, considering different tariff structures and investment costs of both PV systems and batteries. Nicholls et al. [50] found that the inclusion of a battery-based storage system in a rooftop PV system could effectively reduce the cost pay-back time for residential buildings in Queensland, Australia, and South Australia, but a PV system without a storage system is most suited for Tasmania, Australia.

Except for storing electricity generated by renewable sources, batteries are widely used in EVs to eliminate tailpipe emissions [51]. In 2015, Manzetti and Mariasiu [52] reviewed batteries that are suitable for electric vehicles, such as lead acid, nickel-cadmium, nickel-metal-hydride, lithium-ion, lithium-ion polymer, and sodium nickel chloride batteries. Among them, lithium-ion batteries are widely used in EVs due to their higher voltage and density [53].

Furthermore, batteries not only store electricity but also thermal energy. For instance, a thermophysical battery could store and release thermal energy based on the mechanisms of adsorption-desorption and evaporation-condensation [54]. Tarish et al. [55] evaluated the performance (e.g., cooling capacity, coefficient of performance, and specific cooling power) of an adsorption thermophysical battery (ATB) when applied to air conditioning. They analyzed the effect of the adsorbent materials and adsorber bed design on the performance of the ATB for air conditioning applications [56]. A detailed summarization of ATB improvement methods can be found in Tarish et al. [57], which reviewed 158 papers published before 2020, focusing on ATB improvement.

6.2.1.3 Building-integrated thermal energy storage system

The building-integrated thermal energy storage (BITES) system takes advantage of the thermal mass of building envelopes, such as concrete floors or walls. For instance, exterior walls with high thermal mass have been reported to have sufficient energy storage ability to shift peak load from daytime to night [58]. BITES systems could be categorized as active BITES and passive BITES based on whether they are charged by a mechanical force energy input [59]. Passive BITES store/release energy by natural heat transfer, while active BITES charge/discharge the energy storage medium by a mechanical force source. Passive BITES have the advantage of having no maintenance cost, while active BITES usually makes the charge/discharge process faster and more controllable.

To enhance the peak shifting capacity of a building whose construction is mostly made of high thermal mass material (e.g., concrete), Robillart et al. [60] developed a model predictive controller (MPC) to optimize the heating schedule. Liu et al. [61] reviewed macro-encapsulated PCMs that could be integrated into building envelopes to achieve latent energy storage. Sun et al. [62] proposed a heat extraction system to transfer heat from places with sufficient energy storage capacity, such as electrically heated floors (EHFs), to places that are heated by traditional electric baseboards and do not have sufficient heat storage capacity. EHFs refer to heating systems that bury electric wires in concrete slabs; thus they are active TES systems. Thieblemont et al. [63] proposed a self-learning predictive control approach to schedule the heating set-point temperature for a building heated by the EHF. Further modifications and investigations of the self-learning controller could be found in Ref. [64,65].

6.2.1.4 Underground/underwater energy storage system

Underground/underwater ESSs refer to energy storage systems that are placed underground/underwater to reduce the use of onshore land.

Generally, the underground ESS stores heat in the ground during the summer and releases it during the winter [66]. Besides working as an ESS, it is also a system that could effectively use geothermal energy, which is a huge renewable energy source.

The earth-to-air heat exchanger (EAHE) has been widely used to exchange energy between the indoor environment and the earth [67−69]. Liu et al. [70] proposed a vertical EAHE system that requires a small area of land and uses the geothermal energy from deep soil more efficiently. A numerical model of the vertical EAHE was developed in Ref. [71] to conduct a parametric study that investigates the proper design parameters for tube, insulation, and mass flow rates under different soil types. To reduce the temperature fluctuation of air imported into the indoor environment from the tube buried underground, Liu et al. [72] integrated an annular PCM with the vertical EAHE system. Furthermore, Liu et al. [73] used a tubular PCM to make the outlet air temperature of the vertical EAHE system more stable.

Karaca et al. [74] proposed an underwater compressed air ESS to store excess renewable energy by pumping compressed air into underwater balloons. The stored energy could be released to generate electricity using gas turbines. A parametric study of the proposed system found it could provide 365 GWh of electrical energy annually.

6.2.2 Artificial intelligence technologies

AI technologies could be utilized during the design stage or operation stage to optimize the design parameters or operation schedules of ESSs in buildings. Based on the function of AI technologies, this section introduces commonly used AI-based prediction methods, optimization technologies, and control strategies.

6.2.2.1 Artificial intelligence-based prediction methods

AI-based prediction methods usually refer to supervised ML algorithms that use labeled training datasets to explore the mapping function from input variables/attributes to output variables/attributes. Since the fundamental principle of these methods is to find statistical patterns from training datasets, they are also called data-driven prediction methods.

Commonly used supervised ML algorithms include DT, SVM, ANN, DL, and ensemble methods. A detailed description of these methods can be found in Sun et al. [75]. Table 6.1 provides a simple summary of the methods.

Table 6.1 Summary of commonly used supervised machine learning algorithms.

Machine learning algorithm	Concepts	Advantages	Disadvantages
DT	DT is a tree-like model that splits input data from a root node into different internal nodes (subsets) or leaf nodes (outputs) by branches (classification rules) [76].	• Easy to understand and interpret.	• Sensitive to small changes in data. • Cannot determine smooth boundaries.
SVM	SVM model maximizes the margin between different categories during training and then makes predictions based on which side of the margin the data falls [77].	• Effective in high-dimensional spaces. • Memory efficient.	• Negative effect of a large training dataset on its predictive performance. • Cannot perform well when target classes are overlapping.
ANN	ANN is inspired by the biological neural network [78]. It contains neurons in an input layer, several hidden layers, and an output layer.	• The ability to deal with nonlinear predictive problems.	• Longer training time for large neural networks.
DL	DL algorithms are more complex ANNs. It mainly includes three categories: CNN, DNN, and RNN.	• A DNN model requires less neurons than a simple ANN model for the same complex problem.	• DNN models may be overfitted. • DL algorithms can be compute intensive.

(Continued)

Table 6.1 (Continued)

Machine learning algorithm	Concepts	Advantages	Disadvantages
Ensemble methods	Ensemble methods combine the output of multiple machine learning algorithms [79]. They can be classified as bagging, boosting, or stacking models.	• CNN could be effective in avoiding overfitting. • RNN could consider the temporal dynamic of time-series data. • DL algorithms show higher accuracy than ANN. • Higher accuracy than a single data-driven model.	• Compute intensive.

ANN, Artificial neural network; *CNN*, convolutional neural networks; *DL*, deep learning; *DNN*, deep neural networks; *DT*, decision tree; *RNN*, recurrent neural networks; *SVM*, support vector machine.

6.2.2.2 Artificial intelligence-based optimization technologies

This section introduces commonly used swarm intelligence optimization technologies, such as GA, PSO, DE, and ACO. Besides, a brief description is given for other AI-based optimization techniques, such as SA, IA, and AFSA. Note that all these AI-based optimization algorithms are inspired by human brains or animal behaviors, and they are derivative free.

GA is an optimization algorithm inspired by Charles Darwin's theory of natural evolution [80]. Its general procedure is summarized by Sun et al. [81] and presented in Fig. 6.3. It mainly contains five phases:

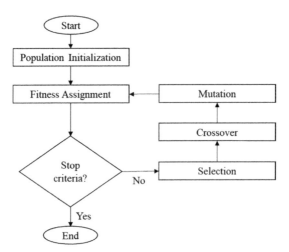

Figure 6.3 General procedure of a GA [81]. *GA*, Genetic algorithm.

population initialization; fitness assignment, selection, crossover, and mutation. The first step is to initialize a population that contains a set of individuals that refers to candidate solutions to the optimization problem. Then, in the fitness assignment step, a fitness function is used to determine how fit each individual is. The fittest individuals are selected to process the crossover and mutation procedures to generate new offspring. The optimization loop terminates if the new generation of the population has converged. GA could be utilized to solve both unconstrained and constrained optimization problems [82].

6.2.2.2.1 Particle swarm optimization

PSO is a meta-heuristic optimization algorithm inspired by the swarm behavior of fish and bird schooling [83]: a set of individuals (also called particles) move around in the search space to find the best solution. The major advantage of PSO is its fewer parameters to tune, while it is limited by its slow convergence speed and poor optimization results when dealing with high dimensional, complex, and large datasets [84].

6.2.2.2.2 Differential evolution

DE iteratively improves the candidate solution based on the evolutionary process until it gets the optimal solution [85]. It could be used to solve unconstrained optimization problems and constrained optimization problems [86]. The differences between DE and GA are: (1) the mutation step is before crossover in DE; (2) DE picks a set of parents in the selection step, while GA takes the two fittest solutions; (3) in the mutation step, DE creates a unit creator based on the target vector and the differential vector, while GA produces new offspring based on the mutation probability.

6.2.2.2.3 Ant colony optimization
ACO is inspired by the foraging behavior of ant colonies [87,88]. The fundamental principle of ACO is to find the best path on a weighted graph that is transformed from the original optimization problem. Note that ACO may suffer an issue with converge speed and solution accuracy when handling a large volume of data [89].

6.2.2.2.4 Simulated annealing
SA is an optimization algorithm that mimics the physical annealing process that heats up a material to an annealing temperature and then cools it down slowly to get the desired structure with the minimum energy configuration [90]. SA could solve both unconstrained and bound-constrained optimization problems [91].

6.2.2.2.5 Immune algorithm
IA that imitates the defense process of the immune system is a kind of heuristic algorithm [92]. It has been proven to be more useful than GA when searching for the optimal solution for some numerical functions [93].

6.2.2.2.6 Artificial fish swarm algorithm
AFSA is an optimization algorithm inspired by fish movement behaviors for collecting food and other various social behaviors (such as immigration) [94]. It shows the advantages of high flexibility, fast convergence speed, and insensitivity to initial settings [95]. A review of AFSA in terms of recent advances and applications is given by Pourpanah et al. [96] in the year 2022.

6.2.2.3 Artificial intelligence-based control strategies
AI-based control strategies could be categorized as predictive-based controllers and predictive-free controllers. A predictive-based controller schedules the operation signals of target systems/devices based on the predicted status of the corresponding systems/devices or buildings. If the predictive-based controller makes predictions based on control signals and integrates the predictive results into the objective function(s) of an optimization problem to get the optimal control signal, it is called an MPC.

A commonly used AI-based predictive-free controller is RL. RL improves the control behavior through trial-and-error interactions with a dynamic environment where the controlled systems/devices are located [97].

Additionally, the fuzzy-logic control system is a widely used machine control method. It is based on fuzzy logic, a reasoning method that resembles human reasoning [98]. Therefore fuzzy-logic control strategies could be classified as a type of AI-based control technique [99]. Yagar [100] and Dubois and Prade [101] illustrate the relationship between fuzzy logic and AI. Kolokotsa [102] reviewed the application of fuzzy logic in building engineering in terms of regulation and modeling of indoor environments and HVAC systems.

6.3 Study cases on the application of artificial intelligence on energy storage systems in buildings

6.3.1 Application of artificial intelligence techniques in the prediction of energy storage systems

ML algorithms could be used to predict the energy storage states and energy storage capacity of ESS. Lv et al. [103] reviewed the applications of ML algorithms on a lithium–ion battery to predict battery materials and battery states. Abualigah et al. [104] summarized commonly used ML technologies for renewable energy systems with ESS. They evaluated the ML methods in terms of robustness, efficiency, accuracy, and generalizability. Based on the analysis, they mentioned that hybrid ML methods usually show higher accuracy than a single method, because hybrid methods could benefit from two or more technologies.

Afram et al. [105] developed a data-driven model to predict the tank temperature. The TES tank was heated by a ground source heat pump and was used to supply hot water to a radiant floor heating system in a heating zone during the heat season. The predicted tank temperature fitted well with the measured data.

Meng et al. [106] developed an Elman neural network (ENN) prediction model to predict the cooling load of an air conditioning system and the charge/discharge duration of an energy storage tank. The ENN was trained based on simulation data collected from a TRNSYS model. Its predictive results are representative of the actual load curve and charge/discharge duration of the energy storage tank. Thus the predictive results are considered in a control strategy for the demand response of an official building to reduce energy consumption and operation costs.

Kumar et al. [107] developed an ANN model to predict the virtual energy storage capacity of a virtual ESS, that is, aggregated refrigerators in residential buildings. Vijayalakshmi et al. [108,109] considered the air conditioner as a virtual ESS and developed a stochastic gradient descent-based ANN model to predict its energy storage capacity. The developed ANN model enabled virtual energy storage for the aggregated air conditioners without any negative effect on thermal comfort. Later, they also developed an ensemble learning model integrated by support vector regression and ANN to predict the virtual energy storage capacity of aggregated air conditioners. Through a comparison with the minute-based smart meter data, the predicted result was validated with high accuracy.

6.3.2 Application of artificial intelligence techniques for optimization and control of energy storage systems

Lee et al. [110] proposed an AI-based MPC scheme to control the charge and discharge rate of TES systems with the aim of minimizing the operation cost. In the MPC scheme, ANN was used to predict the TES system behavior, while the εDE-RJ optimization solver was selected to solve the optimization problem. The proposed MPC was implemented to control a sensible TES tank in an experimental system that included a chiller, pumps, and heat exchangers. The study results show that the AI-based MPC reduced the operational cost by 9.1%−14.6%.

Niu et al. [111] developed an autoregressive model with exogenous inputs for a factory building with a thermal mass storage system and a battery ESS, and then they integrated the validated model into a mixed integer linear model to optimize the energy dispatch with the aim of operational cost saving. The study results show that using a battery ESS could decrease the operational cost by 5.3%, while building thermal storage could further decrease the cost by 4.0%. The cost-saving potential of these ESSs could be preserved, even if adding a constraint for power feeder stability.

Zhang et al. [112] reviewed data-driven MPC and RL algorithms, and their application in building energy management. The challenges of these approaches were summarized and discussed to provide a basis for control method selection. Besides, Brandi et al. [113] compared the cost-saving potential of three AI-based controllers, that is, online-trained deep RL, offline-trained deep RL, and an MPC (based on long short-term memory) network model, when controlling the charge and discharge rate of a cold-water energy storage tank linked to an office building and a chiller. Fig. 6.4 presents the mathematical framework of these three controllers. The results

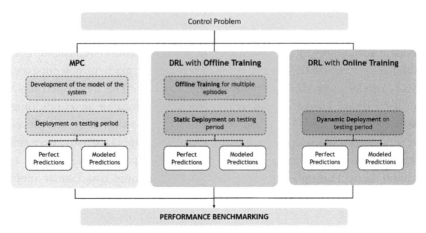

Figure 6.4 Comparison framework of RL and MPC [113]. *MPC,* Model predictive controller; *RL,* reinforcement learning.

show that after a four-week adjustment period with poor control performance, the online-trained deep RL controller shows nearly optimal cost-saving capability: the cooling cost of the online-trained deep RL is only 3.6% higher than the MPC of the last month. However, the online-trained RL shows the advantages of no modeling requirement.

Henze and Schoenmann [114] applied a model-free RL to control the charge/discharge of a thermal storage tank in commercial buildings. The RL is learned through improving the feedback (i.e., monetary cost) for the control actions from the environment. The simulation result shows that RL outperforms conventional control strategies but underperforms predictive-based optimal control. Desportes et al. [115] used a deep RL approach to learn a control policy for a hybrid ESS in a partially islanded building. The hybrid ESS contains a hydrogen storage system and a lead battery. It is powered by PV panels. The control objective is to minimize carbon emissions while ensuring 35% of self-sufficiency. Zhou et al. [116] used an RL algorithm to optimally control the charge/discharge of an ESS in a microgrid based on the predicted PV power generation and load demand of a modified DL model. This study found that RL could effectively reduce the solution time by 61.17% with an acceptable reduction of solution accuracy at 3.13% compared with mixed-integer linear programming. Furthermore, Mason and Grijalva [117] reviewed studies on RL algorithms and their application in building energy management. This literature also outlined the main challenges and potential future research directions for RL.

6.4 Discussion and future works

In the previous section, applications of AI technologies during the operation phase of ESS in buildings have been reviewed with sufficient data collected during the operation phase. In recent years, AI technologies in the building design process have started drawing attention. For instance, Toosi et al. [118] utilized ML models to enhance a lift cycle sustainability assessment process to optimize the ESS size during the design phase. ML models were trained based on simulation results. Besides, He et al. [119] developed an ANN model for a solar space heating system based on the simulation results from TRNSYS. They integrated the ANN model into single objective optimization and multioptimization problems to get the optimal solar collector area and the optimal volume of a water TES tank. The optimization problems are solved by a GA solver. In summary, as the ESS parameter designs based on AI technologies are generally based on simulated data, improving the training data quality and reliability would be an efficient way to make AI technologies more applicable during the design phase.

Sharing electricity storage systems at the community level or district level could enhance the flexibility of energy storage and reduce the levelized costs of storage compared with energy storage in separate units/ buildings [120]. Besides, Syed et al. [121] found that a shared energy microgrid that consists of a PV system and a battery system could improve the overall self-sufficiency of the microgrid by 75%. However, managing energy storage and distribution in a fair way becomes a challenge. To solve this problem, AlSkaif et al. [122] proposed a reputation-based centralized energy management framework to jointly schedule the power consumption of household appliances and fairness-aware energy distribution among different households. The simulation study proved that applying the proposed framework to microgrids could significantly decrease the power imported from the main grid while guaranteeing fairness in energy allocation. Fairness concepts and their possible application in the building engineering domain have been summarized [123]. To achieve fair prediction in different conditions for data-driven predictors, Sun et al. [123] proposed four preprocessing methods to remove discrimination from training datasets before training data-driven building models. The generalizability of these methods on different buildings is evaluated by Sun et al. [124]. Later, Sun et al. [81] proposed four in-processing methods to improve the predictive fairness of regressive models. However, these fairness-aware data-driven methods have not been applied to buildings/ grids equipped with ESSs. It would be a novel and interesting topic to develop fairness-aware data-driven-based energy management frameworks to guarantee fair energy allocation through effective utilization of ESSs.

Another limitation of existing studies on the application of AI to ESS in buildings is that most of the AI-based controllers were tested based on a numerical study, while the experimental application in real buildings is still lacking. For instance, Wang and Hong [125] reported that although RL has drawn increasing attention to building controls, its application in real buildings is still uncommon (accounting for only 11% of all reviewed studies). It is caused by three reasons: (1) Collecting a training dataset and the process of training an RL are time-consuming; (2) Control security and reliability cannot be guaranteed; (3) Generalizability of the developed model needs to be validated. Therefore improving the applicability of AI-based control strategies in real buildings could be valuable for future works.

6.5 Conclusion

In recent years, ESSs has been widely used to store renewable energy and to achieve peak shifting and peak shaving for buildings. ESS contributes in solving the mismatch between supply and demand and easing the burden on electricity grids.

In this chapter, the ESS in buildings, such as water tanks, batteries, building-integrated TES systems, and underground/underwater ESSs, are summarized. Besides, AI technologies have been introduced in terms of AI-based prediction methods, optimization methods, and control strategies. The existing case studies regarding the applications of AI to ESS in buildings for predicting states, optimizing system size operation schedules, and improving control performances have been reviewed.

Based on the discussion on existing studies, future work could focus on ensuring the reliability of training data, considering fairness in prediction and control, and implementing AI-based ESS control strategies in real experiments.

References

[1] Zhou Y, Zheng S, Liu Z, Wen T, Ding Z, Yan J, et al. Passive and active phase change materials integrated building energy systems with advanced machine-learning based climate-adaptive designs, intelligent operations, uncertainty-based analysis and optimisations: A state-of-the-art review. Renewable and Sustainable Energy Reviews 2020;130:109889.

[2] Zhou Y, Liu Z. A cross-scale 'material-component-system' framework for transition towards zero-carbon buildings and districts with low, medium and high-temperature phase change materials. Sustainable Cities and Society 2023;89:104378.

[3] Buildings — Topics, IEA. (n.d.). https://www.iea.org/topics/buildings (accessed November 27, 2022).

[4] Rate Flex D | Hydro-Québec, (n.d.). http://www.hydroquebec.com/residentiel/espace-clients/tarifs/tarif-flex-d.html (accessed March 12, 2020).

[5] Uddin M, Romlie MF, Abdullah MF, Abd Halim S, Abu Bakar AH, Chia Kwang T. A review on peak load shaving strategies. Renewable and Sustainable Energy Reviews 2018;82:3323—32. Available from: https://doi.org/10.1016/j.rser.2017.10.056.

[6] He Y, Zhou Y, Liu J, Liu Z, Zhang G. An inter-city energy migration framework for regional energy balance through daily commuting fuel-cell vehicles. Applied Energy 2022;324:119714.

[7] He Y, Zhou Y, Yuan J, Liu Z, Wang Z, Zhang G. Transformation towards a carbon-neutral residential community with hydrogen economy and advanced energy management strategies. Energy Conversion and Management 2021;249:114834.

[8] Zhou Y, Liu Z, Zheng S. 15 - Influence of novel PCM-based strategies on building cooling performance. Eco-efficient Materials for Reducing Cooling Needs in Buildings and Construction. Woodhead Publishing Series in Civil and Structural Engineering; 2021, p. 329—53.

[9] Lee C-C, Zhang J, Hou S. The impact of regional renewable energy development on environmental sustainability in China. Resources Policy 2023;80:103245. Available from: https://doi.org/10.1016/j.resourpol.2022.103245.

[10] Liu Z, Zhou Y, Yan J, Marcos T. Frontier ocean thermal/power and solar PV systems for transformation towards net-zero communities. Energy 2023;284:128362.

[11] Liu Z, Yu C, Qian K, Huang R, You K, Henk V, et al. Incentive initiatives on energy-efficient renovation of existing buildings towards carbon—neutral blueprints in China: Advancements, challenges and prospects. Energy and Buildings 2023;296:113343.

[12] Liu Z, Xie M, Zhou Y, He Y, Zhang L, Zhang G, et al. A state-of-the-art review on shallow geothermal ventilation systems with thermal performance enhancement system classifications, advanced technologies and applications. Energy and Built Environment 2021. Available from: https://doi.org/10.1016/j.enbenv.2021.10.003.

[13] Weschenfelder F, de Novaes Pires Leite G, Araújo da Costa AC, de Castro Vilela O, Ribeiro CM, Villa Ochoa AA, et al. A review on the complementarity between grid-connected solar and wind power systems. Journal of Cleaner Production 2020;257:120617. Available from: https://doi.org/10.1016/j.jclepro.2020.120617.

[14] Eltawil MA, Zhao Z. Grid-connected photovoltaic power systems: technical and potential problems—a review. Renewable and Sustainable Energy Reviews 2010;14:112—29.

[15] Marocco P, Novo R, Lanzini A, Mattiazzo G, Santarelli M. Towards 100% renewable energy systems: the role of hydrogen and batteries. Journal of Energy Storage 2023;57:106306. Available from: https://doi.org/10.1016/j.est.2022.106306.

[16] Classification of energy storage technologies: an overview, Emerg. Technol. News. (n.d.). https://etn.news/energy-storage/classification-of-energy-storage-technologies-an-overview (accessed January 23, 2023).

[17] Guney MS, Tepe Y. Classification and assessment of energy storage systems. Renewable and Sustainable Energy Reviews 2017;75:1187—97. Available from: https://doi.org/10.1016/j.rser.2016.11.102.

[18] Wilberforce T, Thompson J, Olabi A-G. Classification of energy storage materials. In: Olabi A-G, editor. Encyclopedia of Smart Materials. Oxford: Elsevier; 2022, p. 8—14. Available from: https://doi.org/10.1016/B978-0-12-s803581-8.11762-X.

[19] Hannan MA, Hoque MM, Mohamed A, Ayob A. Review of energy storage systems for electric vehicle applications: issues and challenges. Renewable and Sustainable Energy Reviews 2017;69:771−89. Available from: https://doi.org/10.1016/j.rser.2016.11.171.

[20] Hemmati R, Saboori H. Stochastic optimal battery storage sizing and scheduling in home energy management systems equipped with solar photovoltaic panels. Energy Building 2017;152:290−300. Available from: https://doi.org/10.1016/j.enbuild.2017.07.043.

[21] Thermal performance of the building envelope integrated with phase change material for thermal energy storage: an updated review. Sustainable Cities and Society 2022;79:103690. Available from: https://doi.org/10.1016/j.scs.2022.103690.

[22] Olsthoorn D, Haghighat F, Mirzaei PA. Integration of storage and renewable energy into district heating systems: a review of modelling and optimization. Solar Energy 2016;136:49−64. Available from: https://doi.org/10.1016/j.solener.2016.06.054.

[23] Olsthoorn D, Haghighat F, Moreau A, Lacroix G. Abilities and limitations of thermal mass activation for thermalcomfort, peak shifting and shaving: a review. Building and Environment 2017;118:113−27.

[24] Thieblemont H, Haghighat F, Ooka R, Moreau A. Predictive control strategies based on weather forecast in buildings with energy storage system: a review of the state-of-the art. Energy Building 2017;153:485−500.

[25] Xie M, Qiu Y, Liang Y, Zhou Y, Liu Z, Guo Q. Policies, applications, barriers and future trends of building information modeling technology for building sustainability and informatization in China. Energy Reports 2022;8:7107−26.

[26] Li B, Cheng F, Cai H, Zhang X, Cai W. A semi-supervised approach to fault detection and diagnosis for building HVAC systems based on the modified generative adversarial network. Energy Building 2021;246:111044. Available from: https://doi.org/10.1016/j.enbuild.2021.111044.

[27] Song G, Ai Z, Liu Z, Zhang G. A systematic literature review on smart and personalized ventilation using CO2 concentration monitoring and control. Energy Reports 2022;8:7523−36.

[28] Liu Z, Zhang X, Sun Y, Zhou Y. Advanced controls on energy reliability, flexibility, and occupant-centric control for smart and energy-efficient buildings. Energy and Buildings 2023;113436.

[29] Foucquier A, Robert S, Suard F, Stéphan L, Jay A. State of the art in building modelling and energy performances prediction: a review. Renewable and Sustainable Energy Reviews 2013;23:272−88. Available from: https://doi.org/10.1016/j.rser.2013.03.004.

[30] Amasyali K, El-Gohary NM. A review of data-driven building energy consumption prediction studies. Renewable and Sustainable Energy Reviews 2018;81:1192−205. Available from: https://doi.org/10.1016/j.rser.2017.04.095.

[31] Bourdeau M, Qiang Zhai X, Nefzaoui E, Guo X, Chatellier P. Modeling and forecasting building energy consumption: a review of data-driven techniques, Sustainable Cities and Society 2019;48:101533.

[32] Mohandes SR, Zhang X, Mahdiyar A. A comprehensive review on the application of artificial neural networks in building energy analysis. Neurocomputing 2019;340:55−75. Available from: https://doi.org/10.1016/j.neucom.2019.02.040.

[33] Deb C, Zhang F, Yang J, Lee SE, Shah KW. A review on time series forecasting techniques for building energy consumption. Renewable and Sustainable Energy Reviews 2017;74:902−24. Available from: https://doi.org/10.1016/j.rser.2017.02.085.

[34] Gao T, Lu W. Machine learning toward advanced energy storage devices and systems. IScience 2021;24:101936. Available from: https://doi.org/10.1016/j.isci.2020.101936.

[35] Energy Storage Systems | EMA Singapore, (n.d.). https://www.ema.gov.sg/Energy_Storage_Programme.aspx (accessed January 23, 2023).

[36] Krishan O, Suhag S. An updated review of energy storage systems: classification and applications in distributed generation power systems incorporating renewable energy resources. International Journal of Energy Research 2019;43:6171−210. Available from: https://doi.org/10.1002/er.4285.

[37] Johannes K, Fraisse G, Achard G, Rusaouën G. Comparison of solar water tank storage modelling solutions. Solar Energy 2005;79:216−18. Available from: https://doi.org/10.1016/j.solener.2004.11.007.

[38] Novo AV, Bayon JR, Castro-Fresno D, Rodriguez-Hernandez J. Review of seasonal heat storage in large basins: water tanks and gravel−water pits. Applied Energy 2010;87:390−7. Available from: https://doi.org/10.1016/j.apenergy.2009.06.033.

[39] Dahash A, Ochs F, Janetti MB, Streicher W. Advances in seasonal thermal energy storage for solar district heating applications: a critical review on large-scale hot-water tank and pit thermal energy storage systems. Applied Energy 2019;239:296−315. Available from: https://doi.org/10.1016/j.apenergy.2019.01.189.

[40] Li H, Hou J, Tian Z, Hong T, Nord N, Rohde D. Optimize heat prosumers' economic performance under current heating price models by using water tank thermal energy storage. Energy. 2022;239:122103. Available from: https://doi.org/10.1016/j.energy.2021.122103.

[41] Lyu W, Wang Z, Li X, Deng G, Xu Z, Li H, et al. Influence of the water tank size and air source heat pump size on the energy saving potential of the energy storage heating system. Journal of Energy Storage 2022;55:105542. Available from: https://doi.org/10.1016/j.est.2022.105542.

[42] Zeng C, Cao X, Haghighat F, Yuan Y, Klimes L, Mankibi MEI, et al. Buried water-phase change material storage for load shifting: a parametric study. Energy Building 2020;227:110428. Available from: https://doi.org/10.1016/j.enbuild.2020.110428.

[43] Zeng C, Yuan Y, Cao X, Dardir M, Panchabikesan K, Ji W, et al. Operating performance of multi-modular water-phase change material tanks for emergency cooling in an underground shelter. International Journal of Energy Research 2022;46:4609−29. Available from: https://doi.org/10.1002/er.7454.

[44] Zeng C, Yuan Y, Haghighat F, Panchabikesan K, Cao X, Yang L, et al. Thermo-economic analysis of geothermal heat pump system integrated with multi-modular water-phase change material tanks for underground space cooling applications. Journal of Energy Storage 2022;45:103726. Available from: https://doi.org/10.1016/j.est.2021.103726.

[45] McKenna P, Turner WJN, Finn DP. Thermal energy storage using phase change material: analysis of partial tank charging and discharging on system performance in a building cooling application. Applied Thermal Engineering 2021;198:117437. Available from: https://doi.org/10.1016/j.applthermaleng.2021.117437.

[46] Zhang C, Wei Y-L, Cao P-F, Lin M-C. Energy storage system: current studies on batteries and power condition system. Renewable and Sustainable Energy Reviews 2018;82:3091−106. Available from: https://doi.org/10.1016/j.rser.2017.10.030.

[47] Leng Z, Yuan Y, Cao X, Zeng C, Zhong W, Gao B. Heat pipe/phase change material thermal management of Li-ion power battery packs: a numerical study on coupled heat transfer performance. Energy. 2022;240:122754. Available from: https://doi.org/10.1016/j.energy.2021.122754.

[48] Leng Z, Yuan Y, Cao X, Zhong W, Zeng C. Heat pipe/phase change material coupled thermal management in Li-ion battery packs: optimization and energy-saving assessment. Applied Thermal Engineering 2022;208:118211. Available from: https://doi.org/10.1016/j.applthermaleng.2022.118211.

[49] Akter MN, Mahmud MA, Oo AMT. Comprehensive economic evaluations of a residential building with solar photovoltaic and battery energy storage systems: an

Australian case study. Energy Building 2017;138:332−46. Available from: https://doi.org/10.1016/j.enbuild.2016.12.065.

[50] Nicholls A, Sharma R, Saha TK. Financial and environmental analysis of rooftop photovoltaic installations with battery storage in Australia. Applied Energy 2015;159: 252−64. Available from: https://doi.org/10.1016/j.apenergy.2015.08.052.

[51] Ellingsen LA-W, Majeau-Bettez G, Singh B, Srivastava AK, Valøen LO, Strømman AH. Life cycle assessment of a lithium-ion battery vehicle pack. Journal of Industrial Ecology 2014;18:113−24. Available from: https://doi.org/10.1111/jiec.12072.

[52] Manzetti S, Mariasiu F. Electric vehicle battery technologies: from present state to future systems. Renewable and Sustainable Energy Reviews 2015;51:1004−12. Available from: https://doi.org/10.1016/j.rser.2015.07.010.

[53] Ramkumar MS, Reddy CSR, Ramakrishnan A, Raja K, Pushpa S, Jose S, et al. Review on Li-Ion battery with battery management system in electrical vehicle. Advances in Materials Science and Engineering 2022;2022:e3379574. Available from: https://doi.org/10.1155/2022/3379574.

[54] Narayanan S, Kim H, Umans A, Yang S, Li X, Schiffres SN, et al. A thermophysical battery for storage-based climate control. Applied Energy 2017;189:31−43. Available from: https://doi.org/10.1016/j.apenergy.2016.12.003.

[55] Tarish AL, Khalifa AHN, Hamad AJ. Performance evaluation of adsorption thermophysical battery for air conditioning applications. Thermal Science and Engineering Progress 2022;28:101060. Available from: https://doi.org/10.1016/j.tsep.2021.101060.

[56] Tarish AL, Khalifa AHN, Hamad AJ. Impact of the adsorbent materials and adsorber bed design on performance of the adsorption thermophysical battery: experimental study, case study. Thermal Engineering 2022;31:101808. Available from: https://doi.org/10.1016/j.csite.2022.101808.

[57] Tarish AL, Khalifa AHN, Hamad AJ. Methods of improving the performance of adsorption thermophysical battery based on the operating conditions and structure: a review. IOP Conference Series: Materials Science and Engineering 2020;928:022040. Available from: https://doi.org/10.1088/1757-899X/928/2/022040.

[58] Zhu L, Hurt R, Correia D, Boehm R. Detailed energy saving performance analyses on thermal mass walls demonstrated in a zero energy house. Energy Building 2009;41:303−10. Available from: https://doi.org/10.1016/j.enbuild.2008.10.003.

[59] Sun Y. Heat Extraction System for Augmenting the Heating and Peak Shifting Ability of Electrically Heated Floor Residential Buildings, PhD Thesis, Concordia University. 2018.

[60] Robillart M, Schalbart P, Chaplais F, Peuportier B. Model reduction and model predictive control of energy-efficient buildings for electrical heating load shifting. Journal of Process Control 2019;74:23−34. Available from: https://doi.org/10.1016/j.jprocont.2018.03.007.

[61] Liu Z, Yu ZJ, Yang T, Qin D, Li S, Zhang G, et al. A review on macro-encapsulated phase change material for building envelope applications. Building and Environment 2018;144:281−94.

[62] Sun Y, Panchabikesan K, Joybari MM, Olsthoorn D, Moreau A, Robichaud M, et al. Enhancement in peak shifting and shaving potential of electrically heated floor residential buildings using heat extraction system. Journal of Energy Storage 2018;18:435−46.

[63] Thieblemont H, Haghighat F, Moreau A, Lacroix G. Control of electrically heated floor for building load management: a simplified self-learning predictive control approach. Energy Building 2018;172:442−58. Available from: https://doi.org/10.1016/j.enbuild.2018.04.042.

[64] Luo JT, Joybari MM, Panchabikesan K, Sun Y, Haghighat F, Moreau A, et al. Performance of a self-learning predictive controller for peak shifting in a building integrated with energy storage. Sustainable Cities and Society 2020;60:102285.

[65] Sun Y, Panchabikesan K, Haghighat F, Luo JT, Moreau A, Robichaud M. Development of advanced controllers to extend the peak shifting possibilities in the residential buildings. Journal of Building Engineering 2021;43:103026.

[66] Liu Z, Zeng C, Zhou Y, Xing C. The main utilization forms and current developmental status of geothermal energy for building cooling/heating in developing countries. Utilization of Thermal Potential of Abandoned Wells. Academic Press; 2022, p. 159—90.

[67] Qin D, Liu Z, Zhou Y, Yan Z, Chen D, Zhang G. Dynamic performance of a novel air-soil heat exchanger coupling with diversified energy storage components—modelling development, experimental verification, parametrical design and robust operation. Renewable Energy 2021;167:542—57.

[68] Liu J, Yu Z, Liu Z, Qin D, Zhou J, Zhang G. Performance Analysis of Earth-air Heat Exchangers in Hot Summer and Cold Winter Areas. Procedia Engineering 2017;205:1672—7.

[69] Tang L, Liu Z, Zhou Y, Qin D, Zhang G. Study on a Dynamic Numerical Model of an Underground Air Tunnel System for Cooling Applications—Experimental Validation and Multidimensional Parametrical Analysis. Energies 2020;13:1236.

[70] Liu Z, Yu ZJ, Yang T, Li S, El Mankibi M, Roccamena L, et al. Experimental investigation of a vertical earth-to-air heat exchanger system. Energy Conversion and Management 2019;183:241—51.

[71] Liu Z, (Jerry) Yu Z, Yang T, Roccamena L, Sun P, Li S, et al. Numerical modeling and parametric study of a vertical earth-to-air heat exchanger system. Energy. 2019;172:220—31. Available from: https://doi.org/10.1016/j.energy.2019.01.098.

[72] Liu Z, Yu ZJ, Yang T, El Mankibi M, Roccamena L, Sun Y, et al. Experimental and numerical study of a vertical earth-to-air heat exchanger system integrated with annular phase change material. Energy Conversion and Management 2019;186:433—49. Available from: https://doi.org/10.1016/j.enconman.2019.02.069.

[73] Liu Z, Sun P, Li S, Yu ZJ, El Mankibi M, Roccamena L, et al. Enhancing a vertical earth-to-air heat exchanger system using tubular phase change material. Journal of Cleaner Production 2019;237:117763.

[74] Karaca AE, Dincer I, Nitefor M. A new renewable energy system integrated with compressed air energy storage and multistage desalination. Energy. 2023;268:126723. Available from: https://doi.org/10.1016/j.energy.2023.126723.

[75] Sun Y, Haghighat F, Fung BCM. A review of the-state-of-the-art in data-driven approaches for building energy prediction. Energy Building 2020;221:110022. Available from: https://doi.org/10.1016/j.enbuild.2020.110022.

[76] Induction of decision trees | SpringerLink, (n.d.). https://link.springer.com/article/10.1007/BF00116251 (accessed January 28, 2023).

[77] Basak D. Support vector regression. International Journal of Neural Information Processing—Letters and Reviews 2007;11:203—24.

[78] Kleene SC. Representation of events in nerve nets and finite automata, RAND PROJECT AIR FORCE SANTA MONICA CA. 1951.

[79] Opitz D, Maclin R. Popular ensemble methods: an empirical study. Journal of Artificial Intelligence Research 1999;11:169—98. Available from: https://doi.org/10.1613/jair.614.

[80] Whitley D. A genetic algorithm tutorial. Statistics and Computing 1994;4:65—85. Available from: https://doi.org/10.1007/BF00175354.

[81] Sun Y, Fung BC, Haghighat F. In-processing fairness improvement methods for regression data-driven building models: achieving uniform energy prediction. Energy Building 2022;277:112565.

[82] What Is the Genetic Algorithm? - MATLAB & Simulink, (n.d.). https://www.mathworks.com/help/gads/what-is-the-genetic-algorithm.html (accessed June 3, 2022).

[83] Kennedy J, Eberhart R. Particle swarm optimization. Proc. ICNN95 - Int. Conf. Neural Netw 1995;4:1942−8. Available from: https://doi.org/10.1109/ICNN.1995.488968.

[84] Gad AG. Particle swarm optimization algorithm and its applications: a systematic review. Archives of Computational Methods in Engineering 2022;29:2531−61. Available from: https://doi.org/10.1007/s11831-021-09694-4.

[85] Storn R, Price K. Differential evolution − a simple and efficient heuristic for global optimization over continuous spaces. Journal of Global Optimization 1997;11: 341−59. Available from: https://doi.org/10.1023/A:1008202821328.

[86] Georgioudakis M, Plevris V. A comparative study of differential evolution variants in constrained structural optimization. Frontiers in Built Environment 2020;6. Available from: https://www.frontiersin.org/article/10.3389/fbuil.2020.00102 (accessed June 15, 2022).

[87] Dorigo M, Birattari M, Stutzle T. Ant colony optimization. IEEE Computational Intelligence Magazine 2006;1:28−39. Available from: https://doi.org/10.1109/MCI.2006.329691.

[88] Dorigo M, Di Caro G. Ant colony optimization: a new meta-heuristic, in: Proc. 1999 Congr. Evol. Comput.-CEC99 Cat No 99TH8406. 1999, pp. 1470−1477 2. Available from: https://doi.org/10.1109/CEC.1999.782657.

[89] Yu J, You X, Liu S. A heterogeneous guided ant colony algorithm based on space explosion and long−short memory. Applied Soft Computing 2021;113:107991. Available from: https://doi.org/10.1016/j.asoc.2021.107991.

[90] Bertsimas D, Tsitsiklis J. Simulated annealing. Statistical Science 1993;8:10−15. Available from: https://doi.org/10.1214/ss/1177011077.

[91] What Is Simulated Annealing? - MATLAB & Simulink, (n.d.). https://www.mathworks.com/help/gads/what-is-simulated-annealing.html (accessed January 28, 2023).

[92] Chu C-W, Lin M-D, Liu G-F, Sung Y-H. Application of immune algorithms on solving minimum-cost problem of water distribution network. Mathematical and Computer Modelling 2008;48:1888−900. Available from: https://doi.org/10.1016/j.mcm.2008.02.008.

[93] Chun J-S, Jung H-K, Hahn S-Y. A study on comparison of optimization performances between immune algorithm and other heuristic algorithms. IEEE Transactions on Magnetics 1998;34:2972−5. Available from: https://doi.org/10.1109/20.717694.

[94] Lobato FS, Steffen Jr. V. Fish swarm optimization algorithm applied to engineering system design. Latin American Journal of Solids and Structures 2014;11:143−56. Available from: https://doi.org/10.1590/S1679-78252014000100009.

[95] Neshat M, Sepidnam G, Sargolzaei M, Toosi AN. Artificial fish swarm algorithm: a survey of the state-of-the-art, hybridization, combinatorial and indicative applications. Artificial Intelligence Review 2014;42:965−97. Available from: https://doi.org/10.1007/s10462-012-9342-2.

[96] Pourpanah F, Wang R, Lim CP, Wang X-Z, Yazdani D. A review of artificial fish swarm algorithms: recent advances and applications. Artificial Intelligence Review 2022. Available from: https://doi.org/10.1007/s10462-022-10214-4.

[97] Kaelbling LP, Littman ML, Moore AW. Reinforcement learning: a survey. Journal of Artificial Intelligence Research 1996;4:237−85. Available from: https://doi.org/10.1613/jair.301.

[98] What is Fuzzy Logic in AI and What are its Applications?, Edureka. (2019). https://www.edureka.co/blog/fuzzy-logic-ai/ (accessed January 29, 2023).

[99] Liu Z, Sun Y, Xing C, Liu J, He Y, Zhou Y, et al. Artificial intelligence powered large-scale renewable integrations in multi-energy systems for carbon neutrality transition: challenges and future perspectives. Energy AI 2022;10:100195. Available from: https://doi.org/10.1016/j.egyai.2022.100195.

[100] Yager RR. Fuzzy logics and artificial intelligence. Fuzzy Sets and Systems 1997;90:193−8. Available from: https://doi.org/10.1016/S0165-0114(97)00086-9.

[101] Dubois D, Prade H. What does fuzzy logic bring to AI? ACM Computing Surveys 1995;27:328−30. Available from: https://doi.org/10.1145/212094.212115.

[102] Kolokotsa D. Artificial intelligence in buildings: a review of the application of fuzzy logic. Advances in Building Energy Research 2007;1:29−54. Available from: https://doi.org/10.1080/17512549.2007.9687268.

[103] Lv C, Zhou X, Zhong L, Yan C, Srinivasan M, Seh ZW, et al. Machine learning: an advanced platform for materials development and state prediction in lithium-ion batteries. Advanced Materials 2022;34:2101474. Available from: https://doi.org/10.1002/adma.202101474.

[104] Abualigah L, Zitar RA, Almotairi KH, Hussein AM, Abd Elaziz M, Nikoo MR, et al. Wind, solar, and photovoltaic renewable energy systems with and without energy storage optimization: a survey of advanced machine learning and deep learning techniques. Energies. 2022;15:578. Available from: https://doi.org/10.3390/en15020578.

[105] Afram A, Janabi-Sharifi F, Giorgio G. Data-driven modeling of thermal energy storage tank. In: 2014 IEEE 27th Can. Conf. Electr. Comput. Eng. CCECE; 2014, pp. 1−5. Available from: https://doi.org/10.1109/CCECE.2014.6901009.

[106] Meng Q, Xi Y, Ren X, Li H, Jiang L, Yang L. Thermal energy storage air-conditioning demand response control using Elman neural network prediction model. Sustainable Cities and Society 2022;76:103480. Available from: https://doi.org/10.1016/j.scs.2021.103480.

[107] Kumar KN, Vijayakumar K, Kalpesh C. Virtual energy storage capacity estimation using ANN-based kWh modelling of refrigerators. IET Smart Grid 2018;1:31−9. Available from: https://doi.org/10.1049/iet-stg.2018.0001.

[108] Vijayalakshmi K, Vijayakumar K, Nandhakumar K. Prediction of virtual energy storage capacity of the air-conditioner using a stochastic gradient descent based artificial neural network. Electric Power Systems Research 2022;208:107879. Available from: https://doi.org/10.1016/j.epsr.2022.107879.

[109] Vijayalakshmi K, Vijayakumar K, Nandhakumar K. An ensemble learning model for estimating the virtual energy storage capacity of aggregated air-conditioners. Journal of Energy Storage 2023;59:106512. Available from: https://doi.org/10.1016/j.est.2022.106512.

[110] Lee D, Ooka R, Matsuda Y, Ikeda S, Choi W. Experimental analysis of artificial intelligence-based model predictive control for thermal energy storage under different cooling load conditions. Sustainable Cities and Society 2022;79:103700. Available from: https://doi.org/10.1016/j.scs.2022.103700.

[111] Niu J, Tian Z, Lu Y, Zhao H. Flexible dispatch of a building energy system using building thermal storage and battery energy storage. Applied Energy 2019;243:274−87. Available from: https://doi.org/10.1016/j.apenergy.2019.03.187.

[112] Zhang H, Seal S, Wu D, Bouffard F, Boulet B. Building energy management with reinforcement learning and model predictive control: a survey. IEEE Access 2022;10:27853−62. Available from: https://doi.org/10.1109/ACCESS.2022.3156581.

[113] Brandi S, Fiorentini M, Capozzoli A. Comparison of online and offline deep reinforcement learning with model predictive control for thermal energy management.

Automation in Construction 2022;135:104128. Available from: https://doi.org/10.1016/j.autcon.2022.104128.

[114] Henze GP, Schoenmann J. Evaluation of reinforcement learning control for thermal energy storage systems. HVACR Results 2003;9:259−75. Available from: https://doi.org/10.1080/10789669.2003.10391069.

[115] Desportes L, Fijalkow I, Andry P. Deep reinforcement learning for hybrid energy storage systems: balancing lead and hydrogen storage. Energies. 2021;14:4706. Available from: https://doi.org/10.3390/en14154706.

[116] Zhou K, Zhou K, Yang S. Reinforcement learning-based scheduling strategy for energy storage in microgrid. Journal of Energy Storage 2022;51:104379. Available from: https://doi.org/10.1016/j.est.2022.104379.

[117] Mason K, Grijalva S. A review of reinforcement learning for autonomous building energy management. ArXiv190305196. 2019.

[118] Amini Toosi H, Lavagna M, Leonforte F, Del Pero C, Aste N. A novel LCSA-Machine learning based optimization model for sustainable building design—a case study of energy storage systems. Building and Environment 2022;209:108656. Available from: https://doi.org/10.1016/j.buildenv.2021.108656.

[119] He Z, Farooq AS, Guo W, Zhang P. Optimization of the solar space heating system with thermal energy storage using data-driven approach,. Renewable Energy 2022;190:764−76. Available from: https://doi.org/10.1016/j.renene.2022.03.088.

[120] Müller SC, Welpe IM. Sharing electricity storage at the community level: An empirical analysis of potential business models and barriers. Energy Policy 2018;118:492−503. Available from: https://doi.org/10.1016/j.enpol.2018.03.064.

[121] Syed MM, Hansen P, Morrison GM. Performance of a shared solar and battery storage system in an Australian apartment building. Energy Building 2020;225:110321. Available from: https://doi.org/10.1016/j.enbuild.2020.110321.

[122] AlSkaif T, Luna AC, Zapata MG, Guerrero JM, Bellalta B. Reputation-based joint scheduling of households appliances and storage in a microgrid with a shared battery. Energy Building 2017;138:228−39. Available from: https://doi.org/10.1016/j.enbuild.2016.12.050.

[123] Sun Y, Haghighat F, Fung BC. Trade-off between accuracy and fairness of data-driven building and indoor environment models: a comparative study of pre-processing methods. Energy. 2021;122273. Available from: https://doi.org/10.1016/j.energy.2021.122273.

[124] Sun Y, Fung BCM, Haghighat F. The generalizability of pre-processing techniques on the accuracy and fairness of data-driven building models: a case study. Energy Building 2022;268:112204. Available from: https://doi.org/10.1016/j.enbuild.2022.112204.

[125] Wang Z, Hong T. Reinforcement learning for building controls: the opportunities and challenges. Applied Energy 2020;269:115036. Available from: https://doi.org/10.1016/j.apenergy.2020.115036.

CHAPTER 7

Smart grid with energy digitalization

Xiaohan Zhang

Sustainable Energy and Environment Thrust, Function Hub, The Hong Kong University of Science and Technology (Guangzhou), Nansha, Guangdong, P.R. China

7.1 Introduction

Carbon neutrality transition requires the cross-scale modeling and smart grids integrations [1,2]. Smart grids refer to electricity grids integrating with digital technologies, smart sensors, and advanced software, which can implement real-time sensing and dynamic control on the supply and demand side to achieve specific objectives (e.g., the minimum operation cost) [3]. Smart grid technologies can improve grid resilience, reliability, and flexibility with bi-directional energy flow under optimal control and they can overcome the main drawbacks of the conventional grid, such as (1) one-way power flow, (2) constant energy prices under different demands, and (3) in-efficient metering devices [4].

Fig. 7.1 shows the configuration of a building integrated with smart grid. Smart sensors and smart metering devices could digitalize the energy flows in the smart grid by measuring multiple energy-related parameters in real time. They enable the devices in the system to communicate information and execute the control command [5]. Moreover, functions such as real-time load measuring [6], load forecasting [7], fault detection [8], and demand-side management [9] can be realized with the smart metering system.

Smart energy systems with smart grids and flexible equipment, such as advanced HVAC (heating, ventilating, and air-conditioning) systems, lighting equipment, charging piles, storage units and energy sharing [10−13], have received increasing attention in academia. With proper demand-side management approaches, end-users are enabled to consume electricity more cost-effectively by adjusting the flexible load according to the electricity tariff [14]. Moreover, compared with the conventional power grid, the smart

Advances in Digitalization and Machine Learning for Integrated Building-Transportation Energy Systems
DOI: https://doi.org/10.1016/B978-0-443-13177-6.00002-3

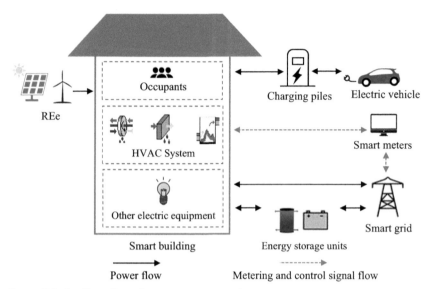

Figure 7.1 Configuration of a smart energy system.

grid shows higher energy efficiency, more controlled carbon emission, and higher renewable energy penetration rate [5].

The penetration of renewable and distributed energy sources introduced various challenges to the energy system, which involve the requirements of: (1) modeling and managing nonlinear and intermittent power systems [15], (2) storing and processing massive amounts of data [5], (3) automatically identifying faults and failures [8], (3) protecting the grid from cyberattack [16], (4) automatically planning and managing the load [17] and so forth. The thriving machine learning (ML) technologies have been considered a powerful tool to address the challenges in the smart grid and promote further development.

7.2 Smart sensor and metering technologies for energy digitalization transformation

The realization of the smart grid in the future necessitates the digitalization of the energy flow in the complex system. The conventional electromechanical watt-hour meter can only transfer energy data in a limited way, which normally requires manual participation [18]. In the smart grid system, the smart sensor and metering technologies are necessary for the operators to automatically track the real-time performance of each component [19], including voltage values, phase angles, and frequency values, as well as perform other functions.

The functions of smart metering technologies consist of data communication and data measurement, as shown in Fig. 7.2. Smart meters can realize bi-directional communication between multiple participants, such as users and providers. Smart meters can also output the control command to home appliances or other inferior devices with a power line or wireless technology [20]. Apart from the communication, measurement, and control modules, smart meters also comprise a logging module that stores consumer information, a timing module that provides timestamps [21], a billing module that can bill the users automatically, and so forth [22].

Compared with traditional electricity meters, smart meters present many advantages in data measurement. To begin with, traditional electricity meters can only measure the consumption of power [18]. However, the smart grid can collect more comprehensive demand-related information, such as the occupant's number inside a building [23], load demand, voltage drops, etc, in order to implement the demand-side management strategy. Another drawback of conventional meters is that they can only measure cumulative energy consumption. However, smart meters can measure more frequently, for example, once per minute, to acquire the

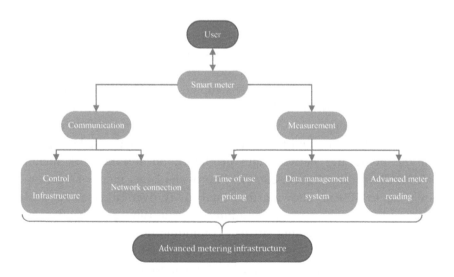

Figure 7.2 Functions and infrastructure of smart meter systems [18].

time series consumption data [18]. Moreover, with the integration of advanced data-driven models in the smart grid system, the collected data needs to be stored in data storage units [24], which is of great help in managing power dispatch and fault detection. The stored data history can also help energy suppliers and consumers check their historical energy consumption [18].

The communication in the smart grid system is bi-directional, performed through wireline and wireless facilities. Wireline communication is preferable under certain conditions because it is more reliable and insensitive to interference compared with wireless communication [21]. The mainstream wireline communication method transfers data over power lines, called power line communication (PLC). According to the different bandwidths, the PLC can be classified into narrowband PLC (NB-PLC) and broadband PLC (BB-PLC) [25]. The NB-PLC operates at 500 kHz frequencies with a bandwidth of up to 500 kbps, which can cover long-distance data transmission of more than 150 km. Resulting from the broad geographical span, the NB-PLC is most suitable for medium-voltage networks [26]. The frequency and bandwidth of the BB-PLC are much higher than those of the NB-PLC, at $2-30$ MHz and 200 Mbps, respectively [25,27]. Common applications of BB-PLC in smart buildings include power quality measurement, distributed generation, and so forth [28]. Apart from the PLC methods, optical methods and digital subscriber line communication are the other two wireline communication methods. The prime advantages of optical methods are that they are capable of transmitting data at high speed and that they can avoid the influence of electromagnetic interference. The digital subscriber line communication methods are based on telephone lines. Therefore part of the infrastructure setup cost for communication can be saved [21]. Compared with the wireline solution, the advantages of wireless communication networks comprise lower installation costs, quicker deployment, and higher flexibility [29,30]. The wireless options for the home area network and neighborhood area network include Wi-Fi, Bluetooth, ZigBee, etc. Wi-Fi provides larger indoor range, typically up to 100 m, allowing for connectivity over larger areas, while Bluetooth offers short-range communication (usually up to 10 m), conserving power and enabling efficient communication between devices in close proximity. And ZigBee technology is cost-effective and consumes very low power, making it suitable for battery-operated devices and applications where power efficiency is crucial [31].

The smart meters need to meet the privacy requirements of multiple participants during operation [32]. The data privacy issue relates to various processes, including data measurement, transfer, storage, etc. The necessary measures including encryption communication protocols, data minimization, and authentication and access control should be developed and applied in the smart grid system [27]. Danezis et al. [33] presented a differential privacy method that could protect consumer privacy by preserving specific data. Gough et al. [34] developed an innovative differential privacy-compliant algorithm to preserve consumer data from smart meters. Results showed that the proposed approach could protect the user's privacy without leading to a significant cost increase. In addition, smart meters also need to allow fault prevention, fault detection, and fault diagnosis in some long-distance transmissions [35], because they can monitor loads of the whole infrastructure [18] through the measurement and communication network. Moreover, with the timestamp of operating data history, the utilities can accurately locate the faults [36].

7.3 Digital technologies for smart energy management in microgrids

7.3.1 Energy flexible buildings for grid power friendliness

In academia, energy flexibility improvement for buildings coupled with demand-side management is considered a valuable measure to balance supply and demand and reduce the peak load of grids. The combination of flexible devices, energy storage equipment, and demand-side management strategy can further improve demand flexibility.

Air-conditioning systems are emphasized in improving demand flexibility in building energy systems due to their high fraction in energy consumption and potential in demand shifting. Utama et al. [37] investigated the demand-side flexibility potential of buildings with air-conditioning systems. Results show that demand-side bidding can help consumers optimize their energy consumption curve and reduce peak demand. Yan et al. [38] proposed a proactive demand response system based on an air-conditioning system. By switching the operation modes, the system realized demand side bidding and demand as a frequency-controlled reserve.

Energy storage units have indispensable values in energy flexibility improvement. By storing the surplus renewable energy and shifting the surplus energy to the grid peak hours, energy storage units can help deal with renewable power fluctuations and achieve peak shaving and valley filling. Thermal storage and battery energy storage are two of the mainstream energy storage units [39]. Both passive and active thermal storages, such as hot water storage tanks, thermal mass of building envelopes, etc., are highly dependent on critical factors of heat losses, such as thermal conductivity, temperature level, area, etc. [40]. Le et al. [40] investigated the characteristics of heat storage and heat conversion under different situations. Results indicated that buildings with different insulation properties are suitable for different strategies. Battery storage has high energy efficiency, and is a useful measure for dealing with the uncertainty of energy system and flattening the demand curve [35]. However, the performance of battery is highly constrained by aging in dynamic power interactions. Zhou et al. [41] proposed a systematic control strategy considering the battery cycle aging, charging/discharging of the battery, time-of-use grid electricity, etc. The depth of battery discharging and the number of cycles were optimized, providing a technical reference for designers in flexibility improvement and battery protection. Zhou et al. [42] established a dynamic self-learning grid-response strategy that improved the relative capacity of the battery on the basis of energy flexibility improvement by controlling the battery charging power during the grid off-peak time. The comparison between multiobjective optimizations and multicriteria decision-making strategies in the study provides a reference for the strategic choice of follow-up researchers.

7.3.2 Grid-response controls on distributed energy units

Grid response refers to buildings' reactions to digital signals from the smart grid. During this process, the smart grid determines the control strategies of dynamic prices and incentives to reduce grid overload. Dynamic price strategy is the most commonly used method. To optimize the energy consumption behavior of end users, the electricity price is usually set to be different between peak hours and off-peak hours. Energy-flexible buildings with distributed energy units (such as advanced flexible devices, small-scale energy generation units, energy storage units, and energy conversion units) have a higher potential to respond to the grid signal with a demand-side management strategy to achieve optimal economic benefits.

Based on the dynamic price, Tavakkoli et al. [43] designed an incentive-based load control strategy for HVAC systems in residential buildings. This strategy can adjust the heating and cooling demands according to the electricity tariff to achieve optimal bonuses. Amin et al. [44] proposed the price-based demand response technique to control HVAC thermostat parameters. With the proposed strategy, the system peak demand showed a reduction of 7.19%−26.8%, which is of great use in reducing the peak load of the local grid. Moreover, energy storage for power shifting from off-peak time to peak time is another way of grid response. Kilkki et al. [45] designed a demand response strategy based on the electric heater and thermal storage devices that can defer energy demand to off-peak hours, while ensuring almost similar thermal comfort. Zhou et al. [46] designed a hybrid storage system consisting of a battery and a hydrogen system that can shift off-peak grid power to peak hours according to electricity prices and effectively reduce the operating cost.

The uncertainties caused by renewable energy penetration can lead to a lower renewable energy utilization ratio. However, this can be alleviated by shifting the excessive renewable energy to the peak hours of the grid with storage units. Mirzaei et al. [47] designed a hydrogen storage system that can store the excess power of the wind turbine and release the power through a hydrogen-based gas turbine. Moreover, a price-based demand response strategy was also involved to reduce the system demand at the grid peak hours. The proposed strategy successfully reduced the wind power spillage and the system operating costs.

The EV is an increasingly valuable source of energy flexibility. By using dynamic pricing for the charging/discharging of EVs in charging piles, the EV fleet can serve as energy resources during peak hours and as storage units during off-peak hours. Zhang et al. [48] proposed a dynamic pricing strategy for EV charging piles, considering the impact of EV charging on the grid load. Simulation results indicate that the dynamic pricing strategy minimized the difference between the peak and valley loads of the grid and improved the economic benefits of the power producer. Li et al. [49] designed a strategy to dynamically adjust the electricity tariff according to the EV demand. Results showed that this strategy could help avoid charging congestion.

In the process of grid response, multiple stakeholders (including the buildings and the grid aggregator) are involved. It is necessary to balance the benefits of each player and incentivize their willingness to participate. As is well known, game theories are powerful methods to achieve equilibrium between multiple players in complex systems. In the demand response scheme, there are typically two kinds of players. The smart grid serves as a leader that gives dynamic prices to the end-user, while smart buildings play the follower role that adjust the real-time demand

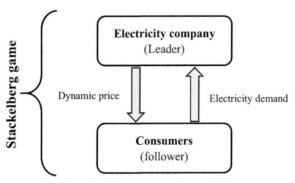

Figure 7.3 Framework of the Stackelberg game theory [52].

according to the electricity price. The Stackelberg game theory, as shown in Fig. 7.3, with a win-win strategy in leader-follower scenarios, is of great use for solving the demand response problem. Nekouei et al. [50] modeled demand response at two levels, that is, the electricity market level and the consumer level. Results of a case study indicated that the Stackelberg game theory can maximize market profit in peak hours. Tang et al. [51] established a grid-building interaction model, and balanced multiple requirements of the smart grid and the building by using Stackelberg game theory. The grid maximized its profit and achieved peak demand reduction, while the building avoided using electricity in the peak period to minimize the operating cost.

7.3.3 Model predictive controls

Model predictive control (MPC) is an advanced algorithm for process control based on iterative, finite-horizon optimization, it is widely used for demand-side management. Fig. 7.4 shows the calculation process of MPC. The first step of performing MPCs is to establish a model to represent the behavior of the controlled system. The predictive model could be the white-box model (mechanism-based model), the black-box model (ML model, statistical model, etc.), or the gray-box model (hybrid model) [53].

The appropriate model should be selected based on factors such as the complexity of the system, available data, control requirements, and computational resources. Alexander et al. [54] established a detailed physics-based model of the HVAC system and the building thermal dynamic model for heating and cooling control. Biyik et al. [55] mathematically developed the thermal zone model, the human comfort model, the battery state of charge model, and the photovoltaics model. These detailed models were applied

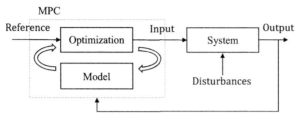

Figure 7.4 Calculation process of model predictive control algorithm.

to the controller design. Mechanism-based models are more explainable and can be calibrated precisely. However, they tend to be computationally expensive for complex systems compared with the black-box model. Manjarres et al. [56] established a regression model of a tertiary building for the predictive control of the HVAC system. The model was trained with the sensor data of the building and could estimate the indoor temperature of the following day. Within the MPC system, the optimizer determined the optimal schedule of the HVAC system according to the predicted temperature. The ML models are more effective since the statistical models tend to be less accurate when dealing with large amounts of data from the building [57]. Megahed et al. [58] proposed an artificial neural network (ANN) model of the system, in which the model can not only estimate the system disturbance but also learn the occupants' behavior. With the neural-network-based MPC, the system can control the indoor temperature efficiently. The gray-box model combines physical knowledge and statistical data [59]. Macarulla et al. [60] evaluated the performance of the gray-box model for ventilation air change rate estimation. The results proved the robust estimation of the gray-box model. Hazyuk et al. [61] first established the low-order model of the building based on physics and then identified the parameters of the model with the least squares method. The proposed model could identify the linear and nonlinear characteristics of the building and permit linear temperature controller.

According to the objective function, the MPC for demand-side management can be further classified. The objective function represents the parameters the designer wants to optimize, such as operating costs and load fluctuations. Hu et al. [62] developed an advanced control strategy based on the MPC, which considered multiple factors, such as weather, dynamic electricity prices, behaviors of the occupants, etc. Results showed that MPC could help comprehensively achieve the optimal strategy that enables peak-time demand shifting, thermal comfort improvement, and operating cost reduction. Razmara et al. [63] presented a real–time MPC control framework, regulating the power imported from the grid, the energy storage units, and the HVAC system in commercial buildings. Results showed that the demand response based on the MPC reduced the operating cost without affecting the thermal comfort of the indoor environment. The objective function of the MPC system [55] is peak power demand reduction with constraints of building dynamics, battery state of charge, and renewable energy generation. Results showed that the MPC significantly reduced the peak load. Toub et al. [64] designed an MPC controller, optimizing energy consumption and building system energy costs. Results showed that the MPC reduced 37% of energy consumption and 70% of operating costs compared with the rule–based controller.

7.4 Role of machine learning in energy digitalization

7.4.1 Machine learning for short-term and long-term dynamic state predictions

ML algorithms for dynamic state prediction within the smart grid are receiving much attention from researchers. The application of machine learning in the smart grid includes load and energy price forecasting, enabling better energy conversion and management.

According to different purposes, the span of load prediction is different. The prediction span of one hour to one week is usually identified as a short–term prediction, which can provide a reference to improve HVAC system performance, demand–side and supply–side management, abnormal energy usage identification, etc. [65]. Yun et al. [66] proposed an autoregressive (AR) model with an exogenous model that can predict the thermal demand 1 hour in advance. The model was compared with the US Department of Energy simulation program to assess accuracy. Results showed that the proposed model was accurate enough for the 1 hour ahead thermal load prediction. Escriva-Escriva et al. [67] presented an ANN model to predict the overall power consumption in buildings. The prediction results served as the baseline for the end users to exert demand–side management. To maximize profits, the energy suppliers need to comprehensively consider factors such as long–term maintenance scheduling, dispatch of energy flow, etc., which require long–term load predictions with a prediction span of more than one year. The long–term prediction is of great help for planning energy resources and provides a reference for the design of the power grid [68] and building energy systems [69,70]. AR, support vector regression, and neural network models are commonly used for pre-diction with high accuracy. Luca et al. [71] presented a support vector regression-based model for long-term energy load prediction. The prediction accuracy of the model was validated with experimental data. Ahmad et al. [72] proposed models to predict the short-, medium-, and long-term energy demand of buildings. The yearly prediction was used to assist in the strategy-making process. The study also compared the performance of different algorithms in performance prediction. Results showed that the binary decision tree has the highest accuracy.

Different from load forecasting, the majority of studies in price forecasting, such as residential electricity prices, pricing schemes, etc. [73–75], are day-ahead, which can be used to achieve the optimal operating strategy for the suppliers and the consumers. Lu et al. [76] proposed a model for day-ahead electricity price prediction. The forecasting data was used to determine the incentive rates for consumers. Huang et al. [77] presented an hour-ahead prediction model for reducing energy costs. This model proved to be capable of working for real-time price-based demand response.

7.4.2 Machine learning for dynamic power dispatch and controls

From the perspective of the consumer, ML applications for optimal power dispatch and control in the smart grid context typically comprise the adjusting of heat pumps, water heaters, etc. [78,79]. Reinforcement learning (RL) excels in dynamic energy systems by adapting control strategies in real time to handle fluctuations in demand and renewable energy sources. Its model-free nature allows it to learn directly from data, making it adaptable to complex and changing power system dynamics without relying on precise models [75,76]. Xiong et al. [80] conducted a hardware-in-loop simulation to compare the performance of different control strategies in the context of a hybrid energy storage system. Results showed that the RL methods outperformed the other methods. The RL method has great adaptability to the new application environment, which makes it more suitable for the real world [81]. Ruelens et al. [78] utilized batch RL for demand response, which can achieve the optimal heat pump control strategy without expert knowledge. EV charging stations are becoming important components of flexible loads in the smart grid. Hafez et al. [82] developed a neural network model to determine the optimal operation of EV charging stations. The objective function of the model considered the benefits of the local distributors and the EV owners.

ML algorithms are also commonly used to solve the optimization problem. In a study conducted by Yoon et al. [83], an ANN-based model was used to calculate the optimal electricity price, balancing the benefits of the distribution system operators and the building operators on the basis of ensuring occupants' energy demand. Claessens et al. [84] employed the convolutional neural network (CNN) for direct load control. Results showed that the CNN model performs well in reducing the electricity cost of buildings.

7.4.3 Machine learning for optimal scheduling and operation

There is a large range of components in the smart grid system that can be used for demand response. This makes it necessary to automatically schedule and operate these devices. As the demand response problem can be decomposed into optimization subproblems, ML methods (such as genetic algorithms, Q-learning) can be used to solve them. The algorithms at the aggregator's level and the consumer's level differ in scale and scope, leading to different requirements for the scheduling strategy. The models that work for the aggregators are required to be more scalable compared with the consumer-level model [85]. To achieve the maximum benefits for the aggregators, Medved et al. [86] proposed an intelligent aggregator scheduling and control scheme with approximate Q-learning methods. The study results showed that the proposed method showed minimal scheduling violations compared with the economic scheduling approach and the energy scheduling approach. The objectives of scheduling and operating problems at the consumer level are typically to minimize the electricity cost [87] and energy consumption [79]. Lin et al. [88] employed a nondominated sorting genetic algorithm for the in-home power scheduling of domestic appliances and proved its feasibility with an experiment. Hafez et al. [82] proposed that ML models can also adjust the charging schedule in the EV charging context. Pedrasa et al. [89] scheduled the day–ahead demand response load with the binary particle swarm algorithm. With the constraint of total load curtailment requirements, the objectives are to minimize total cost, ensure essential operation requirements of interruptible loads, and reduce interruption frequency.

7.5 Future smart grid with energy digitalization and artificial intelligence

The design of the future smart grid is focused on providing reliable, flexible, and cost-effective services to consumers [90], and improving the environmental performance of the energy system. In the future, with smart sensors and smart meters serving as the foundation of the smart grid, the energy system can predict dynamic state parameters. Moreover, energy suppliers and consumers will be capable of matching supply and demand, managing renewable integration, and facilitating demand response with a wider range of accurate power data and bi-directional communication within the smart metering system. It is envisioned that advanced algorithms, such as artificial intelligence, can facilitate many functions of the smart grid and improve the performance of power management with ML-based system state and user behavior forecasting [91,92]. Despite the advantages and advances of ML strategies, challenges still need to be further studied, such as data processing methods for decision-making and lightweight ML solutions. Moreover, given the fact that the existing demand-side management networks are usually based on cloud servers with high latency [93], further research on edge computing is necessary. Multi-energy integration and energy-efficient operation will promote the low-carbon transition [94].

References

[1] Zhou Y, Siqian Zheng, Jiachen Lei, Yunlong Zi. A cross-scale modelling and decarbonisation quantification approach for navigating Carbon Neutrality Pathways in China. Energy Conversion and Management 2023;297:117733.

[2] Zhou Y. Worldwide carbon neutrality transition? Energy efficiency, renewable, carbon trading and advanced energy policies. Energy Reviews 2023;2(2):100026.

[3] Baidya S, Potdar V, Pratim RP, Nandi C. Reviewing the opportunities, challenges, and future directions for the digitalization of energy. Energy Research & Social Science 2021;81.

[4] Manoj KN. Blockchain: enabling wide range of services in distributed energy system. Beni-Suef University Journal of Basic Applied Sciences 2018;701−4.

[5] Ibrahim MS, Dong W, Yang Q. Machine learning driven smart electric power systems: current trends and new perspectives. Applied Energy 2020;272.

[6] Berges ME, Goldman E, Matthews HS, Soibelman L. Enhancing electricity audits in residential buildings with nonintrusive load monitoring. Journal of Industrial Ecology 2010;14:844.

[7] Ertugrul OF. Forecasting electricity load by a novel recurrent extreme learning machines approach. International Journal of Electrical Power & Energy Systems 2016;78:429−35.

[8] Liu WY, Tang BP, Han JG, Lu XN, Hu NN, He ZZ. The structure healthy condition monitoring and fault diagnosis methods in wind turbines: a review. Renewable & Sustainable Energy Reviews 2015;44:466−72.

[9] Li D, Jayaweera SK. Machine-learning aided optimal customer decisions for an interactive smart grid. IEEE Systems Journal 2015;9:1529−40.

[10] Zhou Y. Incentivising multi-stakeholders' proactivity and market vitality for spatiotemporal microgrids in Guangzhou-Shenzhen-Hong Kong Bay Area. Applied Energy 2022;120196.

[11] Zhou Y. Energy sharing and trading on a novel spatiotemporal energy network in Guangdong-Hong Kong-Macao Greater Bay Area. Applied Energy 2022;119131.

[12] Zhou Y. Sustainable energy sharing districts with electrochemical battery degradation in design, planning, operation and multi-objective optimisation. Renewable Energy 2023;202:1324−41.

[13] Zhou Y, Peter D. Lund. Peer-to-peer energy sharing and trading of renewable energy in smart communities — trading pricing models, decision-making and agent-based collaboration. Renewable Energy 2023;207:177−93.

[14] Yuce B, Rezgui Y, Mourshed M. ANN-GA smart appliance scheduling for optimised energy management in the domestic sector. Energy and Buildings 2016;111:311−25.

[15] Shi H, Xu MH, Li R. Deep learning for household load forecasting—a novel pooling deep RNN. IEEE Transactions on Smart Grid 2018;9:5271−80.

[16] Ozay M, Esnaola I, Vural FTY, Kulkarni SR, Poor HV. Machine learning methods for attack detection in the smart grid. IEEE Transactions on Neural Networks and Learning Systems 2016;27:1773−86.

[17] Palensky P, Dietrich D. Demand side management: demand response, intelligent energy systems, and smart loads. IEEE Transactions on Industrial Informatics 2011;7:381−8.

[18] Avancini DB, Rodrigues J, Martins SGB, Rabelo RAL, Al-Muhtadi J, Solic P. Energy meters evolution in smart grids: a review. Journal of Cleaner Production 2019;217:702−15.

[19] Granderson J. Energy information handbook: applications for energy-efficient building operations. 2011.

[20] Liu Z, Xiang Zhang, Ying Sun, Yuekuan Zhou. Advanced controls on energy reliability, flexibility and occupant-centric control for smart and energy-efficient buildings. Energy and Buildings 2023;297:113436.

[21] Kabalci Y. A survey on smart metering and smart grid communication. Renewable and Sustainable Energy Reviews 2016;57:302−18.

[22] Yang Z, Chen YX, Li Y-F, Zio E, Kang R. Smart electricity meter reliability prediction based on accelerated degradation testing and modeling. International Journal of Electrical Power & Energy Systems 2014;56:209−19.

[23] Neksa PRH, Zakeri GR, Schieffoe PA. CO_2 heat pump water heating: characteristics, system design and experimental results. In: Neksa P, Rekstad H, Zakeri GR, Schieffoe PA, editors. International Journal of Refrigeration. 1998, p. 171−8.

[24] Hou WG, Ning ZL, Guo L, Zhang X. Temporal, functional and spatial big data computing framework for large-scale smart grid. IEEE Transactions on Emerging Topics in Computing 2019;7:369−79.

[25] Ancillotti E, Bruno R, Conti M. The role of communication systems in smart grids: architectures, technical solutions and research challenges. Computer Communications 2013;36:1665−97.

[26] Papadopoulos TA, Kaloudas CG, Chrysochos AI, Papagiannis GK. Application of narrowband power-line communication in medium-voltage smart distribution grids. IEEE Transactions on Power Delivery 2013;28:981−8.

[27] Galli S, Scaglione A, Wang Z. For the grid and through the grid: the role of power line communications in the smart grid. Proceedings of the IEEE 2011;99:998−1027.

[28] Mlýnek P, Rusz M, Benešl L, Sláčik J, Musil P. Possibilities of broadband power line communications for smart home and smart building applications. Sensors. 2021;21:240.

[29] Lee J-S, Su Y-W, Shen C-C. A comparative study of wireless protocols: Bluetooth, UWB, ZigBee, and Wi-Fi. IECON 2007-33rd Annual Conference of the IEEE Industrial Electronics Society: IEEE; 2007, p. 46−51.

[30] Rafiei M, Elmi SM, Zare A. Wireless communication protocols for smart metering applications in power distribution networks. In: 2012 Proceedings of 17th Conference on Electrical Power Distribution: IEEE; 2012, p. 1−5.

[31] Mahmood A, Javaid N, Razzaq S. A review of wireless communications for smart grid. Renewable and Sustainable Energy Reviews 2015;41:248−60.

[32] Jawurek M, Johns M, Kerschbaum F. Plug-in privacy for smart metering billing. Privacy Enhancing Technologies: 11th International Symposium, PETS 2011, Waterloo, ON, Canada, July 27—29, 2011 Proceedings 11: Springer; 2011, p. 192—210.

[33] Danezis G, Kohlweiss M, Rial A. Differentially private billing with rebates. Information Hiding: 13th International Conference, IH 2011, Prague, Czech Republic, May 18—20, 2011, Revised Selected Papers 13: Springer; 2011. p. 148—62.

[34] Gough MB, Santos SF, AlSkaif T, Javadi MS, Castro R, Catalão JPS. Preserving privacy of smart meter data in a smart grid environment. IEEE Transactions on Industrial Informatics 2021;18:707—18.

[35] Muscas C, Pau M, Pegoraro PA, Sulis S. Smart electric energy measurements in power distribution grids. IEEE Instrumentation & Measurement Magazine 2015;18:17—21.

[36] Fadel E, Gungor VC, Nassef L, Akkari N, Maik MGA, Almasri S, et al. A survey on wireless sensor networks for smart grid. Computer Communications 2015;71:22—33.

[37] Utama C, Troitzsch S, Thakur J. Demand-side flexibility and demand-side bidding for flexible loads in air-conditioned buildings. Applied Energy 2021;285.

[38] Yan CC, Xue X, Wang SW, Cui BR. A novel air-conditioning system for proactive power demand response to smart grid. Energy Conversion and Management 2015;102:239—46.

[39] Niu JD, Tian Z, Lu YK, Zhao HF. Flexible dispatch of a building energy system using building thermal storage and battery energy storage. Applied Energy 2019;243:274—87.

[40] Le Dreau J, Heiselberg P. Energy flexibility of residential buildings using short term heat storage in the thermal mass. Energy. 2016;111:991—1002.

[41] Zhou YK, Cao SL, Hensen JLM, Hasan A. Heuristic battery-protective strategy for energy management of an interactive renewables-buildings-vehicles energy sharing network with high energy flexibility. Energy Conversion and Management 2020;214.

[42] Zhou Y. A dynamic self-learning grid-responsive strategy for battery sharing economy—multi-objective optimisation and posteriori multi-criteria decision making. Energy. 2022;126397.

[43] Tavakkoli M, Fattaheian-Dehkordi S, Pourakbari-Kasmaei M, Liski M, Lehtonen M. An incentive based demand response by HVAC systems in residential houses. 2019 IEEE PES Innovative Smart Grid Technologies Europe (ISGT-Europe): IEEE; 2019, p. 1—5.

[44] Amin U, Hossain MJ, Fernandez E. Optimal price based control of HVAC systems in multizone office buildings for demand response. Journal of Cleaner Production 2020;270.

[45] Kilkki O, Alahaivala A, Seilonen I. Optimized control of price-based demand response with electric storage space heating. IEEE Transactions on Industrial Informatics 2015;11:281—8.

[46] Zhou L, Zhou Y. Study on thermo-electric-hydrogen conversion mechanisms and synergistic operation on hydrogen fuel cell and electrochemical battery in energy flexible buildings. Energy Conversion and Management 2023;277:116610.

[47] Mirzaei MA, Yazdankhah AS, Mohammadi-Ivatloo B. Stochastic security-constrained operation of wind and hydrogen energy storage systems integrated with price-based demand response. International Journal of Hydrogen Energy 2019;44:14217—27.

[48] Zhang XP, Liang YN, Zhang YK, Bu YH, Zhang HY. Charge pricing optimization model for private charging piles in Beijing. Sustainability. 2017;9.

[49] Li DS, Zouma A, Liao JT, Yang HT. An energy management strategy with renewable energy and energy storage system for a large electric vehicle charging station. Etransportation 2020;6.

[50] Nekouei E, Alpcan T, Chattopadhyay D. Game-theoretic frameworks for demand response in electricity markets. IEEE Transactions on Smart Grid 2015;6:748—58.

[51] Tang R, Wang SW, Li HX. Game theory based interactive demand side management responding to dynamic pricing in price-based demand response of smart grids. Applied Energy 2019;250:118−30.

[52] Alshehri K, Liu J, Chen XD, Basar T. A game-theoretic framework for multiperiod-multicompany demand response management in the smart grid. IEEE Transactions on Control Systems Technology 2021;29:1019−34.

[53] Mariano-Hernandez D, Hernandez-Callejo L, Zorita-Lamadrid A, Duque-Perez O, Garcia FS. A review of strategies for building energy management system: model predictive control, demand side management, optimization, and fault detect & diagnosis. Journal of Building Engineering 2021;33.

[54] Schirrer A, Brandstetter M, Leobner I, Hauer S, Kozek M. Nonlinear model predictive control for a heating and cooling system of a low-energy office building. Energy and Buildings 2016;125:86−98.

[55] Biyik E, Kahraman A. A predictive control strategy for optimal management of peak load, thermal comfort, energy storage and renewables in multi-zone buildings. Journal of Building Engineering 2019;25.

[56] Manjarres D, Mera A, Perea E, Lejarazu A, Gil-Lopez S. An energy-efficient predictive control for HVAC systems applied to tertiary buildings based on regression techniques. Energy and Buildings 2017;152:409−17.

[57] Fan C, Xiao F, Yan CC, Liu CL, Li ZD, Wang JY. A novel methodology to explain and evaluate data-driven building energy performance models based on interpretable machine learning. Applied Energy 2019;235:1551−60.

[58] Megahed TF, Abdelkader SM, Zakaria A. Energy management in zero-energy building using neural network predictive control. IEEE Internet of Things Journal. 2019;6:5336−44.

[59] Bacher P, Madsen H. Identifying suitable models for the heat dynamics of buildings. Energy and Buildings 2011;43:1511−22.

[60] Macarulla M, Casals M, Forcada N, Gangolells M, Giretti A. Estimation of a room ventilation air change rate using a stochastic grey-box modelling approach. Measurement 2018;124:539−48.

[61] Hazyuk I, Ghiaus C, Penhouet D. Optimal temperature control of intermittently heated buildings using Model Predictive Control: Part I − Building modeling. Building and Environment 2012;51:379−87.

[62] Hu MM, Xiao F, Jorgensen JB, Li RL. Price-responsive model predictive control of floor heating systems for demand response using building thermal mass. Applied Thermal Engineering 2019;153:316−29.

[63] Razmara M, Bharati GR, Hanover D, Shahbakhti M, Paudyal S, Robinett RD. Building-to-grid predictive power flow control for demand response and demand flexibility programs. Applied Energy 2017;203:128−41.

[64] Toub M, Reddy CR, Razmara M, Shahbakhti M, Robinett RD, Aniba G. Model-based predictive control for optimal MicroCSP operation integrated with building HVAC systems. Energy Conversion and Management 2019;199.

[65] Amasyali K, El-Gohary NM. A review of data-driven building energy consumption prediction studies. Renewable & Sustainable Energy Reviews 2018;81:1192−205.

[66] Yun K, Luck R, Mago PJ, Cho H. Building hourly thermal load prediction using an indexed ARX model. Energy and Buildings 2012;54:225−33.

[67] Escriva-Escriva G, Alvarez-Bel C, Roldan-Blay C, Alcazar-Ortega M. New artificial neural network prediction method for electrical consumption forecasting based on building end-uses. Energy and Buildings 2011;43:3112−19.

[68] Ahmad T, Chen HX. Potential of three variant machine-learning models for forecasting district level medium-term and long-term energy demand in smart grid environment. Energy. 2018;160:1008−20.

[69] Catalina T, Iordache V, Caracaleanu B. Multiple regression model for fast prediction of the heating energy demand. Energy and Buildings 2013;57:302−12.
[70] Naji S, Keivani A, Shamshirband S, Alengaram UJ, Jumaat MZ, Mansor Z, et al. Estimating building energy consumption using extreme learning machine method. Energy. 2016;97:506−16.
[71] Ghelardoni L, Ghio A, Anguita D. Energy load forecasting using empirical mode decomposition and support vector regression. IEEE Transactions on Smart Grid 2013;4:549−56.
[72] Ahmad T, Chen HX, Huang RG, Guo YB, Wang JY, Shair J, et al. Supervised based machine learning models for short, medium and long-term energy prediction in distinct building environment. Energy. 2018;158:17−32.
[73] Giovanelli C, Liu X, Sierla S, Vyatkin V, Ichise R, IEEE. Towards an Aggregator that Exploits Big Data to Bid on Frequency Containment Reserve Market. 43rd Annual Conference of the IEEE-Industrial-Electronics-Society (IECON). Beijing, PEOPLES R CHINA; 2017, p. 7514−9.
[74] Pal S, Kumar R, IEEE. Price Prediction Techniques for Residential Demand Response Using Support Vector Regression. 7th IEEE Power India International Conference (PIICON). Bikaner, INDIA; 2016.
[75] Chen QF, Liu N, Wang C, Zhang JH, IEEE. Optimal Power Utilizing Strategy for PV-based EV Charging Stations Considering Real-Time Price. IEEE Transportation Electrification Conference and Expo (ITEC Asia-Pacific). Beijing, PEOPLES R CHINA; 2014.
[76] Lu RZ, Hong SH. Incentive-based demand response for smart grid with reinforcement learning and deep neural network. Applied Energy 2019;236:937−49.
[77] Huang XF, Hong SH, Li YT. Hour-ahead price based energy management scheme for industrial facilities. IEEE Transactions on Industrial Informatics 2017;13:2886−98.
[78] Ruelens F, Claessens BJ, Vandael S, De Schutter B, Babuska R, Belmans R. Residential demand response of thermostatically controlled loads using batch reinforcement learning. IEEE Transactions on Smart Grid 2017;8:2149−59.
[79] Ahmed MS, Mohamed A, Shareef H, Homod RZ, Abd Ali J, IEEE. Artificial Neural Network Based Controller for Home Energy Management Considering Demand Response Events. International Conference on Advances in Electrical, Electronic and Systems Engineering (ICAEES). Putrajaya, MALAYSIA 2016, p. 506−9.
[80] Xiong R, Duan YZ, Cao JY, Yu QQ. Battery and ultracapacitor in-the-loop approach to validate a real-time power management method for an all-climate electric vehicle. Applied Energy 2018;217:153−65.
[81] Du Y, Zandi H, Kotevska O, Kurte K, Munk J, Amasyali K, et al. Intelligent multizone residential HVAC control strategy based on deep reinforcement learning. Applied Energy 2021;281.
[82] Hafez O, Bhattacharya K. Integrating EV Charging Stations as Smart Loads for Demand Response Provisions in Distribution Systems. IEEE Transactions on Smart Grid 2018;9:1096−106.
[83] Yoon AY, Kim YJ, Zakula T, Moon SI. Retail electricity pricing via online-learning of data-driven demand response of HVAC systems. Applied Energy 2020;265.
[84] Claessens BJ, Vrancx P, Ruelens F. Convolutional neural networks for automatic state-time feature extraction in reinforcement learning applied to residential load control. IEEE Transactions on Smart Grid 2018;9:3259−69.
[85] Antonopoulos I, Robu V, Couraud B, Kirli D, Norbu S, Kiprakis A, et al. Artificial intelligence and machine learning approaches to energy demand-side response: a systematic review. Renewable & Sustainable Energy Reviews 2020;130.
[86] Medved T, Artac G, Gubina AF. The use of intelligent aggregator agents for advanced control of demand response. Wiley Interdisciplinary Reviews-Energy and Environment 2018;7.

[87] Bahraini S, Wong VWS, Huang JW. An online learning algorithm for demand response in smart grid. IEEE Transactions on Smart Grid 2018;9:4712−25.

[88] Lin YH, Tsai MS. An advanced home energy management system facilitated by nonintrusive load monitoring with automated multiobjective power scheduling. IEEE Transactions on Smart Grid 2015;6:1839−51.

[89] Pedrasa MAA, Spooner TD, MacGill IF. Scheduling of demand side resources using binary particle swarm optimization. IEEE Transactions on Power Systems 2009;24:1173−81.

[90] Chebbo, Maher. EU SmartGrids Framework Electricity Networks of the future 2020 and beyond. Power Engineering Society General Meeting; 2007, p. 1−8.

[91] Chung YW, Khaki B, Li TY, Chu CC, Gadh R. Ensemble machine learning-based algorithm for electric vehicle user behavior prediction. Applied Energy 2019;254.

[92] Liang TK, Zeng B, Liu JQ, Ye LF, Zou CF. An unsupervised user behavior prediction algorithm based on machine learning and neural network for smart home. IEEE Access 2018;6:49237−47.

[93] Kulkarni S, Gu QC, Myers E, Polepeddi L, Liptak S, Beyah R, et al. Enabling a decentralized smart grid using autonomous edge control devices. IEEE Internet of Things Journal 2019;6:7406−19.

[94] Zhou Y. Low-carbon transition in smart city with sustainable airport energy ecosystems and hydrogen-based renewable-grid-storage-flexibility. Energy Reviews 2022;1 (1):100001.

[95] Mehrjerdi H, Bornapour M, Hemmati R, Ghiasi SMS. Unified energy management and load control in building equipped with wind-solar-battery incorporating electric and hydrogen vehicles under both connected to the grid and islanding modes. Energy. 2019;168:919−30.

[96] Zhou Y, Song A. A hierarchical control with thermal and electrical synergies on battery cycling ageing and energy flexibility in a multi-energy sharing network. Renewable Energy 2023;212:1020−37.

[97] Zhou Y. Artificial intelligence in renewable systems for transformation towards intelligent buildings. Energy and AI 2022;10:100182.

[98] Zhou Y. Advances of machine learning in multi-energy district communities-mechanisms, applications and perspectives. Energy and AI 2022;10:100187.

[99] Zhou Y. Ocean energy applications for coastal communities with artificial intelligencea state-of-the-art review. Energy and AI 2022;10:100189.

CHAPTER 8

Peer-to-peer energy trading with advanced pricing and decision-making mechanisms

Siqian Zheng[1] and Yuekuan Zhou[2,3,4]

[1]Department of Architecture and Civil Engineering, City University of Hong Kong, Kowloon, Hong Kong SAR, P.R. China
[2]Sustainable Energy and Environment Thrust, Function Hub, The Hong Kong University of Science and Technology (Guangzhou), Nansha, Guangdong, P.R. China
[3]Department of Mechanical and Aerospace Engineering, The Hong Kong University of Science and Technology, Clear Water Bay, Hong Kong SAR, P.R. China
[4]Division of Emerging Interdisciplinary Areas, The Hong Kong University of Science and Technology, Clear Water Bay, Hong Kong SAR, P.R. China

8.1 Introduction: peer-to-peer energy sharing

The energy transition to sustainable energy sources is an effective way of mitigating global warming and preventing energy crisis. Many nations are therefore dedicated to using clean energy, including hydropower, wind, solar, and geothermal energy. However, the renewable power supply does not always correspond with the demand of energy users. Utilizing intermittent and unpredictable energy generation in the grid becomes problematic at high levels.

By integrating renewable generators, such as solar panels and wind turbines, energy consumers are becoming prosumers that can produce and consume electricity. Incentive policies have been carried out to remunerate electrical power fed from renewable systems into the utility grid. For instance, net metering and feed-in tariff (FiT) schemes are proposed to encourage onsite renewable installation. In net metering, the excess generation injected into the power grid is monitored and later deducted from the prosumer's electricity usage over a billing period (e.g., a month). Under the FiT scheme, the generation sold to the utility grid can be paid at a fixed export price. In recent years, distributed energy resources (DERs) with grid connections are being rapidly deployed, which imposes both technical and financial pressure on the utilities. Many countries have started to reduce or even remove support and subsidies for renewable

Advances in Digitalization and Machine Learning for Integrated Building-Transportation Energy Systems
DOI: https://doi.org/10.1016/B978-0-443-13177-6.00013-8

generation. Therefore alternative solutions are desired by DER owners to earn profits and contribute to the balance of energy.

Given this context, peer-to-peer (P2P) energy trading has been considered a promising way for managing DERs in community microgrids and providing local electricity markets. Generally, energy trading can be based on large-scale transactions between districts, middle-scale transactions between microgrids, or small-scale transactions between individual buildings. In this chapter, prosumer-based energy trading is described as the energy transactions between prosumers and consumers in a connected community where excess power from the DERs of prosumers can be shared among their neighbors. In a traditional peer-to-grid (P2G) scheme, when a prosumer generates excess power, it would be delivered to the grid or stored in electric storage systems. Considerable advantages of P2P energy sharing have been demonstrated in the literature, compared with P2G systems, including (1) the increase in total and individual cost savings [1], (2) the enhancement of load matching and grid interaction [2], and (3) the reduction of carbon emissions [3].

This chapter presents P2P energy trading from the perspectives of local market design, participant energy management, and market mechanisms, including price determination and decision-making methods.

Fig. 8.1 sketches the main structure for P2P energy systems. The P2P energy system needs a virtual layer and a physical layer to implement energy transactions and management. The virtual layer has a platform for

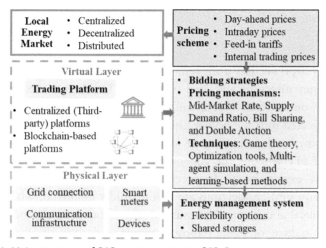

Figure 8.1 Main structure of P2P energy systems. *P2P*, Peer-to-peer.

the participants to make decisions on their trading parameters and ensure the security of the energy supply, while the physical layer facilitates the actual energy dispatch and services for the participants if the transaction in the virtual layer is agreed upon. Tushar et al. [4] have described the essential components of the two layers in detail. The energy trading platforms in the virtual layer can be classified into centralized and decentralized platforms. The P2P energy trading in the centralized platform can be conducted through a third party, e.g., a coordinator or an aggregator. For instance, Zhang et al. [5] proposed a centralized software platform to enable P2P energy trading in a low-voltage microgrid. The energy trading simulation includes bidding, energy exchanging, and settlement processes in sequence. Different from third-party-based centralized platforms, blockchain-based platforms provide a secure environment for direct energy trading among various entities in a decentralized manner [6].

8.2 Peer-to-peer energy trading markets

Based on how the trading process is carried out and how information is shared among the participants, the P2P energy market can be categorized into three groups: centralized market, distributed market, and decentralized market, as illustrated in Fig. 8.2.

8.2.1 Centralized markets

As shown in Fig. 8.2A, in a centralized market, a central controller (i.e., coordinator) communicates with each peer and directly controls the energy shared between peers and the operation of flexible devices within the network. The significant advantage of this market type is that it can

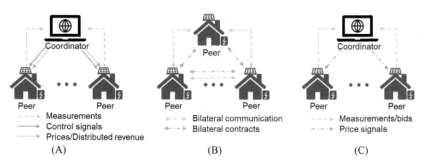

Figure 8.2 Typical categories of markets for energy sharing: (A) centralized, (B) decentralized, and (C) distributed.

be used to find globally optimal solutions, such as maximizing overall social welfare in a cooperative manner.

The coordinator usually distributes the sharing benefits of the entire system following some predefined rules. To encourage the peers to first trade locally, the P2P trading prices are usually set between the FiT prices and the prices for grid-imported electricity (i.e., $\lambda_{FiT} \leq \lambda_{P2P} \leq \lambda_{imp}$). In some studies, electricity is assumed to be traded among peers at a fixed price. For example, in Englberger et al. [7], the local electricity price is kept equal to the average between the electricity retailer's selling and buying prices. This simple approach reduces computation complexity significantly; without considering the fairness of cost distribution.

8.2.1.1 Stability and fair allocation

In a centralized market, a group of players enforces cooperative behaviors to maximize the overall benefit [8]. Two critical issues need to be considered in a cooperative (or coalitional) game. The first issue is how to form an energy coalition, which means the stability of the coalition. The core of a cooperative game is defined as the set of all imputations that ensure the stability of a grand coalition [9], which is implemented to analyze the stability of the market designed in [10]. The second issue is how to allocate the benefits, which is a fairness-related problem.

The concept of Shapley value and nucleolus have been utilized to realize proper allocation for multiple players in P2P energy trading systems [11,12]. The nucleolus gives an allocation solution by minimizing the dissatisfaction of all players and locates in the core if it is nonempty [13]. The Shapley value approach distributes the cost or payoff among the players based on their marginal contributions to the energy-sharing coalition [12]. The coalition is unstable when the Shapley approach is used, but the Shapley value is not in the core. Hence, the stabilizing allocations given by the nucleolus and the fair allocation using the Shapley value provide important baselines to evaluate the performance of other allocation methods.

The approach based on cooperative games can help to guarantee the participants' benefits in centralized energy-sharing systems where operational rules and pricing mechanisms are designed by the system operator. However, a key limitation for the calculation of nucleolus and Shapley value is that the computational complexity increases exponentially with the number of peers [13].

8.2.1.2 Decision-making in central coordinators

In a centralized market, peers have no visibility of other peers throughout the transactions; therefore a central coordinator is responsible for orchestrating the energy sharing between peers. The DERs can be controlled by the coordinator to optimize the overall benefits or provide specific types of ancillary services.

Heuristic rule-based and optimization-based decision-making approaches are widely implemented in centralized sharing systems. Decision-making algorithms based on a set of predefined rules provide simplicity to the system [14]. Optimization algorithms, including linear optimization [15,16], mixed-integer linear programming [17], and meta-heuristic algorithms [18,19], lead to an optimal operation strategy. Statistical decision theory has also been applied to decision-making under conditions of uncertainty and risk [20,21].

The centralized system is vulnerable to single-point failures at the central controller. Communication and computation costs would increase with the number of participants. However, the protection of information privacy [22] and the reduction of market power exploitation due to information asymmetry [23] are becoming important in the energy markets.

8.2.2 Decentralized markets

In a decentralized market, as shown in Fig. 8.2B, peers within the P2P system directly communicate with each other and determine the trading parameters. In other words, peers exchange information and energy according to their own preferences without the requirement of a central intermediary. The participants have full control of the decision-making process to achieve their selfish or altruistic goals. This type of market has better scalability and privacy protection than the centralized market [24]. Different methods have been applied in decentralized markets, including bilateral contract networks [25], consensus approaches [26], and multiagent methods [27,28]. However, the global optimum cannot be easily achieved in a decentralized market, and the efficiency of decentralized systems is relatively lower compared with coordinated systems.

Furthermore, in most existing studies, decentralized markets are designed based on particular beliefs about how players will make decisions [24]. However, the decision-making processes of individual users may be highly complex, uncertain, and diverse from each other, which indicates that the markets designed in practice need careful consideration and performance assessment.

8.2.2.1 Blockchain-enabled energy trading

Blockchain technology has been utilized in the decentralized energy trading market for providing security, privacy, immutability, and transparency thanks to the distinctive features of distributed ledgers, cryptography, consensus algorithms, and smart contracts [26,29]. The distributed ledger technology can record transactions in a permanent and verifiable way, which serves as Bitcoin's underlying fabric. Similar to Bitcoin, a digital currency called NRGcoin has been introduced for transactive energy in smart grids [30].

The blockchain-enabled energy trading framework usually consists of three entities: energy nodes, aggregators, and smart meters [31]. Each energy node (e.g., a building or an electric car) chooses its own role (i.e., buyer, seller, or idle node) according to its state and plans. The buyers would pay the sellers according to the records of energy trading parameters calculated by smart meters in real time. Energy aggregators manage trading-related events and provide communication services for energy nodes. For instance, advanced metering infrastructures can be the aggregators in microgrids.

Although blockchain technology provides opportunities for distributed energy trading systems, its data management, standardization, and incentive mechanisms still need to be improved for widespread implementation [32]. It is necessary to increase scalability while lowering computation and transmission costs without sacrificing security and privacy.

8.2.2.2 Consideration of network and operation constraints

For real applications, there is a need to consider the network constraints and the constraints of DERs. In fully decentralized energy trading systems with no coordination, it is difficult to follow the network's technical requirements, as the total energy that could be traded is uncertain for network operators. The active participation of users without any coordinated control may cause network issues such as overvoltage. Hence, network constraints should be included in the decentralized energy trading models to prevent capacity and voltage problems [33]. Guerrero et al. [34] assessed the effects of P2P transactions on the low-voltage network and proposed a method to guarantee energy exchanges without violating network constraints.

When blockchain is adopted for coordinating DERs in transactive energy markets, the physical constraints of the DER operation should be considered, instead of treating DERs as ideal financial assets. In Münsing et al. [29], blockchain is used to coordinate physical devices with operational constraints and facilitate the aggregator step of decentralized optimization to solve the optimal power flow problem in a P2P energy system.

8.2.3 Distributed markets

In a distributed market, as shown in Fig. 8.2C, the trading process is performed in a decentralized manner while the communication is centralized. Contrary to fully decentralized markets, in distributed markets a coordinator is involved to coordinate the behaviors of peers. Compared with centralized markets, the coordinator in distributed markets collects less information from their peers and influences their decision-making process indirectly by broadcasting pricing signals. Under this circumstance, the participants have a higher level of autonomy and privacy, which provides a compromised solution that combines the advantages of centralized markets and fully decentralized markets.

Indirect demand-side control schemes rely on different pricing mechanisms. P2P participants are offered time-varying prices that reflect the value of electricity in the day and are motivated to decide on their individual behaviors. The design of suitable pricing mechanisms plays a crucial role in distributed markets. A detailed design mechanism has been introduced in Section 8.4.

Table 8.1 presents a summary of different market structures for energy sharing. Different kinds of markets may coexist in one community to balance their merits and limitations.

8.3 Energy management for system participants

Operators and prosumers with DERs are the main participants in energy-sharing markets. The DERs mainly include distributed generation, energy storage systems, and flexible demands, such as electronic devices, electric vehicles (EVs), and HVAC (heating, ventilation, and air conditioning) loads, as depicted in Fig. 8.3.

8.3.1 Distributed multienergy generation

Distributed generation systems include photovoltaic (PV) panels, wind turbines, hydrogen fuel cells, combined heat and power (CHP) plants, etc. Due to the spatiotemporal complementarity of energy profiles, renewable energy sharing is the most popular P2P sharing strategy studied

Table 8.1 Summary of typical market structures for peer-to-peer (P2P) energy trading.

Markets	Merits	Limitations	References
Centralized	Achieves global optimization, taking into account network constraints easily; improves the efficiency of power systems as no bidding process is needed.	Privacy and fairness issues; exponential computational and communication loads caused by the increase in community scale; vulnerable to single-point failures.	[35−37]
Decentralized	Providing good scalability and autonomy; only requiring local peer decision and communication between P2P partners.	Difficult to manage network constraints and DERs' physical constraints and to enhance system efficiency.	[25,38,39]
Distributed	Peers have a high level of autonomy and preserve their privacy by sharing limited information.	Difficult to achieve an optimal solution and design a suitable pricing mechanism regarding both market efficiency and fairness.	[40−42]

DERs, Distributed energy resources.

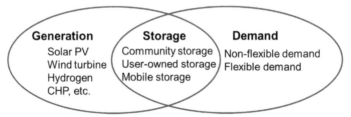

Figure 8.3 Various types of energy resources in prosumer-based energy trading markets.

extensively in academia. The local power-sharing method in microgrids could lower the energy loss on the AC line by 60% without the need for a large energy storage capacity [43]. Prosumers can purchase cheaper energy from each other rather than from the utility grid.

Most existing studies of renewable energy sharing have focused on the participation of PV prosumers [44,45], as residential households are usually equipped with rooftop solar panels. For high-rise buildings, small-scale wind turbines can be integrated to produce onsite energy. However, solar and wind generation are uncontrollable and highly dependent on weather conditions. Many efforts have been made to enhance renewable penetration, improve forecast accuracy [46], and mitigate intermittency [47]. The coordination between renewable energy storage, energy storage systems, and demand response could help mitigate the uncertainty of production and lead to economic benefits [47]. It is indicated that energy sharing is preferred for heterogeneous building groups [48].

The connection between multienergy systems is tighter than before due to cogeneration technologies. Researchers have investigated the opportunities for hydrogen sharing in the P2P energy network [49,50]. A local energy market framework has been developed for trading electricity and hydrogen by Xiao et al. [51], in which participants consist of loads, hydrogen vehicles, renewable generators, and a hydrogen storage system. For sharing energy among multiple microgrids, a multiagent management framework incorporating fuel cell vehicles, CHP units, renewable energy, and energy storage has been proposed, to increase local renewable consumption and enhance economic benefits of microgrids by scheduling the arrangements of fuel cell vehicles [52].

8.3.2 Electrical energy storage sharing

Given the imbalance between energy demand and intermittent generation, prosumers can be integrated with battery energy storage systems. The storage systems provide the necessary flexibility and increase network reliability by maintaining frequency and voltage at the required levels. However, the requirement for large storage capacity leads to high investment costs, and frequent charging/discharging processes cause unnecessary energy losses and shorten battery service life. Thus it is essential to consider the design and operation of storage systems for achieving high renewable penetration and energy arbitrage [53].

In the light of energy sharing microgrids, prosumers can store cheap energy or excess energy in shared storage systems and release the energy when needed. The user could own one shared battery jointly or possess distributed batteries individually. The impact of BESS ownership structures on the energy sharing network was compared by Rodrigues et al. [54]. The results showed that for the studied buildings, the distributed user-owned battery model had a higher net present value compared with the third-party owned community battery. In addition to stationary batteries, EV sharing is a recently emerging approach for sharing stored energy between buildings based on bidirectional charging technology [55]. A novel energy management approach utilizing the building-vehicle-building interaction can realize the energy savings from 38% to 73% [56]. Furthermore, plug-in EVs can extend the energy boundary of the building to improve the energy flexibility. The role and mechanism of battery and hydrogen storages in a district energy community have been reviewed [57].

8.3.3 Demand-side management

The P2P participants may contain energy inflexibilities and flexibilities. Flexible buildings and EVs can respond to incentive signals sent from the grid or third parties and change their energy consumption profiles [58], with the development of digital and smart infrastructure. Multiple prosumers could achieve mutual benefits through cooperative demand-side management in the energy-sharing system by imposing a proper pricing mechanism [59].

The strategic behaviors of energy users in P2P energy trading networks mostly depend on the scheduling and control of flexible demand and storage systems. Building energy loads can be adjusted by proactively controlling the operation of electric devices, lighting, as well as in HVAC systems. Therefore researchers have studied the synergistic effects of different demand-side management approaches to participate in transactive energy systems [7,60]. Sánchez Ramos et al. [61] managed the prosumers with heat pumps and solar panels in an energy-sharing district using different precooling strategies through a central controller, leading to an increased self-sufficiency ratio of 64%. Englberger et al. [7] compared decentralized and central decision-making approaches in P2P systems with stationary and mobile energy storages and indicated that the central approach has the greatest profitability potential. As shown in Fig. 8.4, the

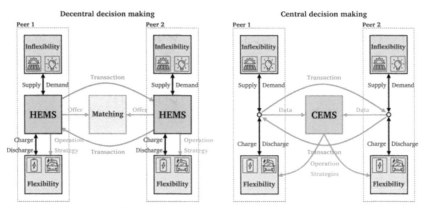

Figure 8.4 Illustration of decentralized (left) and central (right) decision-making in a P2P energy system [7]. The two approaches differ in the type of EMS, market mechanism, data availability, and computation cost. *EMS*, Energy management system; *P2P*, peer-to-peer.

home energy management system at each peer calculates the offers and sends them to the matching platform in the decentralized approach, whereas the community energy management system determines the operation of all peers to yield the optimal techno–economic performance in the centralized approach.

8.4 Advanced technologies for peer-to-peer energy trading implementation

8.4.1 Energy pricing models based on predefined rules

A suitable trading mechanism with reasonable prices, which determines the sharing behaviors and participation willingness of energy users (i.e., prosumers and consumers), is essential for the success of a P2P energy market. P2P trading prices can be computed based on predefined mathematical formulations, which makes the pricing process more straightforward and time-efficient for energy users than other mechanisms (i.e., double auction models, bilateral contract–based mechanisms, and discriminate pricing strategies using game theory).

Table 8.2 summarizes three mechanisms in terms of mathematical formulations for energy selling and buying prices. $P^t_{gen,i}$ and $P^t_{dem,i}$ are the energy supply and demand of peer i at time t, while λ^t_{buy} and λ^t_{sell} denote the import and export prices of the grid, respectively.

Table 8.2 Peer-to-peer pricing based on predefined mathematical formulations.

Mechanism	Calculation
SDR	$$\lambda_{P2P,sell}^t = f(SDR) = \begin{cases} \dfrac{\lambda_{sell}^t \cdot \lambda_{buy}^t}{SDR^t \cdot (\lambda_{buy}^t - \lambda_{sell}^t) + \lambda_{sell}^t}, & 0 \le SDR^t \le 1 \\[2mm] \lambda_{sell}^t, & SDR^t > 1 \end{cases}$$ $$\lambda_{P2P,buy}^t = \begin{cases} \lambda_{P2P,sell}^t \cdot SDR^t + \lambda_{buy}^t \cdot (1 - SDR^t), & 0 \le SDR^t \le 1 \\[2mm] \lambda_{sell}^t, & SDR^t > 1 \end{cases}$$
MMR	$$\lambda_{P2P,sell}^t = \begin{cases} \lambda_{mid}^t = (\lambda_{buy}^t + \lambda_{sell}^t)/2, & \displaystyle\sum_{i=1}^{n} P_{gen,i}^t \le \sum_{i=1}^{n} P_{dem,i}^t \\[3mm] \dfrac{\lambda_{mid}^t \cdot \sum_{i=1}^{n} P_{dem,i}^t + \left(\sum_{i=1}^{n} P_{gen,i}^t - \sum_{i=1}^{n} P_{dem,i}^t\right) \cdot \lambda_{sell}^t}{\sum_{i=1}^{n} P_{gen,i}^t}, & \displaystyle\sum_{i=1}^{n} P_{gen,i}^t > \sum_{i=1}^{n} P_{dem,i}^t \end{cases}$$ $$\lambda_{P2P,buy}^t = \begin{cases} \lambda_{mid}^t = (\lambda_{buy}^t + \lambda_{sell}^t)/2, & \displaystyle\sum_{i=1}^{n} P_{dem,i}^t \le \sum_{i=1}^{n} P_{gen,i}^t \\[3mm] \dfrac{\lambda_{mid}^t \cdot \sum_{i=1}^{n} P_{gen,i}^t + \left(\sum_{i=1}^{n} P_{dem,i}^t - \sum_{i=1}^{n} P_{gen,i}^t\right) \cdot \lambda_{buy}^t}{\sum_{i=1}^{n} P_{dem,i}^t}, & \displaystyle\sum_{i=1}^{n} P_{dem,i}^t > \sum_{i=1}^{n} P_{gen,i}^t \end{cases}$$
BS	$$\lambda_{P2P,sell}^t = \dfrac{\left(\sum_{j=1}^{n} \sum_{i=1}^{T} P_{dem,i}^x - \sum_{j=1}^{n} \sum_{i=1}^{T} P_{dem,i}^x\right) \cdot \lambda_{sell}^t}{\sum_{j=1}^{n} \sum_{i=1}^{T} P_{gen,i}^x}$$ $$\lambda_{P2P,buy}^t = \dfrac{\left(\sum_{j=1}^{n} \sum_{i=1}^{T} P_{dem,i}^x - \sum_{j=1}^{n} \sum_{i=1}^{T} P_{dem,i}^x\right) \cdot \lambda_{buy}^t}{\sum_{j=1}^{n} \sum_{i=1}^{T} P_{dem,i}^x}$$

BS, Bill sharing; *MMR*, mid-market rate; *SDR*, supply–demand ratio.

Fig. 8.5 presents the P2P prices under two predefined mechanisms. The basic principle of economics, that is, the relationship between price and supply—demand ratio (SDR) is inversely proportional, is applied to an SDR pricing mechanism, as shown in Fig. 8.5. Liu et al. [62] formulated an SDR-based mechanism in an energy-sharing model with price-based demand response for the microgrids. In a study by Bogensperger and Ferstl [63], the SDR mechanism was found to be particularly profitable for producers in municipalities with high demand and little supply, while the opposite is true for consumers. Long et al. [35] further modified the SDR pricing mechanism and developed a model using a compensating price to ensure each customer is better off in PV-battery systems.

In a mid-market rate (MMR) pricing mechanism, as shown in Fig. 8.5, the P2P price is set in the middle of the import price and the export price when demand and supply are balanced within the community [64]. Once the energy mismatch exists, the additional costs or revenues of energy exchanges with the grid must be incorporated into the pricing process.

Bill sharing (BS) is a cost-sharing mechanism in which market participants share the energy cost and income evenly over a specified period of time according to the amount of energy consumption and generation [40]; that leads to a constant P2P selling and buying price. This mechanism is more straightforward compared with the SDR and MMR mechanisms. Nevertheless, the energy trading prices can only be determined at the end of the billing period since the cost and revenue can only be computed ex-post.

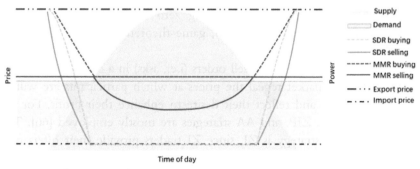

Figure 8.5 Illustration of P2P energy prices under SDR and MMR mechanisms. *MMR*, Mid-market rate; *P2P*, peer-to-peer; *SDR*, supply—demand ratio.

Although peers do not directly participate in the price setting or negotiation process in the predefined pricing mechanisms, they can influence the internal prices by changing their individual energy demand and supply. Zhou et al. [40] compared the above-mentioned energy trading mechanisms based on a multiagent simulation framework, for the flexibility of energy storage and flexible demands. To enhance the convergence of the multiagent simulation, they proposed step length control and learning process involvement techniques during the iterative process. The results show that the SDR-based mechanism is better than the other two mechanisms in their overall performance. According to Fig. 8.5, SDR prices are characterized by the greatest fluctuations, which offers good incentives for additional renewables, flexibility, and storage capacities. From an economic perspective, BS may be unfavorable for producers since revenues are generated and distributed among them only when there is an oversupply; however, the BS method provides the most stability [63].

8.4.2 Bidding strategies in double auction

In a single-sided auction with demand-side or supply-side bidding only, the merit order of the bids received determines the clearing prices [65]. A double auction is more attractive since both buyers and sellers can submit bids simultaneously to match trading partners and trade excess electricity. Offers from participants consist of bid and ask prices along with quantity.

Bidding is the first step in energy trading when P2P participants sign contracts with each other prior to real-time energy sharing. The bidding process can be one-shot or continuous until it reaches equilibrium. Several bidding strategies have been developed to accommodate interactive energy trading markets, including zero intelligence (ZI), ZI plus (ZIP), adaptive aggressiveness (AA), game-theoretic bidding, and intelligent bidding strategies.

Buy orders (i.e., bids) and sell orders (i.e., asks) in a continuous double auction (CDA) market reveal the prices at which participants are willing to accept a trade and reflect their desire to enhance their profit. For the CDA market, ZI, ZIP, and AA strategies are mostly employed [66]. The simplest bidding strategy is ZI, since ZI traders provide their offers uniformly at random prices within a range that may be associated with the grid tariffs, without any previous knowledge about the local market [67]. Compared with ZI traders, ZIP traders use an adaptive mechanism that allows a trader's profit margin to alter over time based on previous orders. The profit margin determines the difference between their limit prices and their asks/bids. Specifically, according to Guerrero et al. [34], ZIP traders are subject to a budget constraint (the maximum and minimum price) that prevents them from buying or selling at a loss. Regarding the frequent price fluctuation, an AA strategy is proposed in the blockchain-based market to adjust the quotation in a timely manner using a learning approach according to the price variation, which makes the AA strategy perform better than ZIP in trading efficiency.

Fig. 8.6 shows the demand and supply curves of quotes in a double-auction based matching process. A transaction happens when the highest price of bids is higher or equal to the lowest price of asks, and the transaction price (i.e., equilibrium price) is the middle point of the two prices [68]. The order book is updated repeatedly until all matches have been finished, and the main grid is used to ensure energy balance. Thus the allocation of energy and the setting of corresponding prices can be automatically implemented. Since the equilibrium price and equilibrium quantity are not known before the transaction actually occurs, Zhang et al. [68] used an iterative double-auction mechanism to maximize equilibrium quantity and social welfare, where buyers and sellers adjust quotes to improve individual profits according to previous transaction volumes and the equilibrium price. It is found that the proposed algorithm increased the average hourly social welfare by 22.3% compared with the ZI strategy. Nevertheless, the performance of the auction-based pricing strategy was evaluated without the participation of any energy storage or elastic demand in the microgrid.

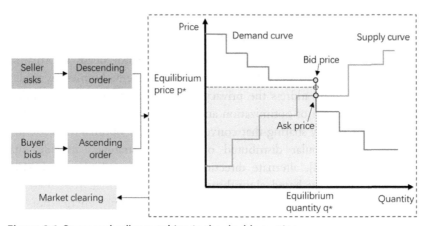

Figure 8.6 Buyer and seller matching in the double auction.

8.4.3 Game-theoretic approach

Game theoretic tools have been extensively applied to the design and operation of modern electricity markets for analyzing complex interactions among independent players [69]. Cooperative games and noncooperative games are the two main categories of games. Noncooperative games are used for the simulation of bidding and are formulated in competitive markets where players make decisions independently [70]. Nash equilibrium can be used to solve the competitive game. If each player has chosen a strategy and no player could benefit from changing their own strategies while others' strategies remain unchanged, the current set of strategies and the associated payoffs constitute a Nash equilibrium.

Examples of noncooperative games used for energy trading are generalized Nash games [41], Stackelberg games [42], and multileader multifollower games [71]. Paudel et al. [42] proposed a game-theoretic model for P2P energy trading in a community microgrid based on noncooperative game, evolutionary game, and Stackelberg game theories. In Opadokun et al. [71], higher priorities are given to the buyers with more energy demand and the sellers offering the lowest price, and then the optimum utilities for both buyers and sellers are evaluated using linear programming.

The game-theoretic bidding strategy provides for the determination of the deviation of lower and higher prices according to previous orders [72]. With the best-offer game-theoretic approach, supply and demand are not considered in the bid-price determination process. This leads to high economic efficiency since all participants try to offer accommodating prices [72].

8.4.4 Distributed optimization

In a centralized energy-sharing market, a coordinator needs information from all the peers to solve the optimization problem. Numerous studies have attempted to address the privacy issue by interpreting energy prices as the dual variable of optimization and implementing distributed iterative methods for energy sharing that converge to socially optimal market equilibrium [73]. Popular distributed optimization strategies include dual decomposition [74], alternate direction method of multipliers (ADMM) [75], and consensus-based algorithms [76]. Baroche et al. [75] used the consensus ADMM to solve the endogenous economic energy dispatch problem for a distributed P2P market.

8.4.5 Multiagent-based simulation

An agent is an autonomous entity that can be considered to perceive its environment through sensors and act upon the environment through actuators, as shown in Fig. 8.7. If the value of any agent is influenced by another agent's behavior, the environment is considered a multiagent environment. A multiagent system (MAS) includes multiple agents that interact, cooperate, and negotiate with each other. It is suitable for simulating the behavior of autonomous prosumers to achieve their individual benefits in decentralized energy trading platforms [77].

Intelligent agents have the capacity to share information and make decisions for utility maximization in specific environments by interacting with other agents. In Celik et al. [78], a MAS is used for modeling smart homes and an aggregator as the agent in the neighborhood. The aggregator agent is the supervisor agent who determines the whole profile and dynamic prices through communication with home agents. Home agents are independent and self-interested decision-makers who aim at maximizing individual welfare while achieving near-optimal performance at Nash equilibrium in a noncooperative coordination game.

The MAS technology can decentralize the highly complex energy system and enable more active user participation [27]. There has been a lot of discussion on the operation of microgrids based on MASs in recent years [79,80]. Decentralized rule-based decision algorithms provide simplicity without heavy computational loads, yet they do not necessarily lead to optimality. However, the agents' continuous interactions in a multiagent framework can result in near-optimal behaviors in energy

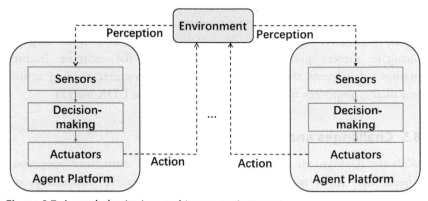

Figure 8.7 Agent behavior in a multiagent environment.

communities. Jiang et al. [81] implemented a negotiation-based MAS to coordinate energy use and storage to maximize the self-consumption of neighborhoods. In addition, a multiagent-based game theory reverse auction model was proposed to enable the competition between DERs within an hour-ahead market to find the cheapest energy supply for a lumped load in a microgrid [82].

8.4.6 Machine learning techniques

Various types of machine learning techniques [83], such as supervised learning in characterizing household energy patterns for demand response [84] and unsupervised learning in personalized retail pricing design [85], have been utilized in energy management systems and electricity markets.

Reinforcement learning (RL) has the capacity to handle energy trading issues and the interaction of direct energy entities with the market environment [79,86,87]. Charbonnier et al. [87] studied different combinations of exploration sources, reward definitions, and multiagent learning frameworks in which coordinated agents learn to control the flexibility provided by EVs, space heating, and other loads. The proposed RL-based strategies address the scalability issue of residential energy flexibility coordination in a cost-efficient and privacy-preserving manner and further contribute to lower energy costs, slower battery depreciation, and less greenhouse gas emissions.

Artificial intelligence algorithms are used for predicting and addressing other data analytic-related issues. While a large-scale energy trading market requires an analysis of massive and diverse data for real-time response to market changes, the combination of blockchain and learning algorithms in artificial intelligence could realize secure and intelligent P2P energy trading [88]. Moreover, the integration of machine learning with model-based approaches helps in learning stochastic environments with less complexity and lower computation load [89]. Chen et al. [90] developed a surrogate market prediction model based on the extreme learning machine that learns the relationship between prosumer bidding actions and market responses from the historical data in the CDA market.

8.5 Challenges and future works

With the deployment of DERs, the conventional centralized operational scheme may suffer from conflicting interests, incentive inadequacy, and privacy concerns. Different forms of P2P energy-sharing markets have their advantages, disadvantages, and applications. Distributed markets and decentralized markets provide effective alternatives to centralized markets due to lower computation and communication costs, greater prosumer autonomy, and the ability to provide privacy protection at some level.

For distributed and decentralized markets, the communication hazard among peers is a potential challenge. Most existing energy trading mechanisms in these markets are based on a synchronous computation that each peer must wait for information from its connected partners; however, long communication delays or communication losses may make the synchronization impracticable [91].

Furthermore, energy trading mechanisms should be properly designed to protect the interests and privacy of all participants. Generally, users are not willing to declare their local information truthfully unless there is an incentive for them. Through a strategy-proof mechanism, market participants would voluntarily disclose their information to achieve the effect of eliminating information asymmetry.

Future studies can pay attention to providing stable P2P energy trading markets. In centralized markets in which prosumers are only considered price takers, stability mainly depends on the allocation rules and profit distribution. Based on cooperative game theory, the Shapley value and the nucleolus have been considered promising methods to allocate the energy bill/income. However, the methods are not computationally efficient if a large number of peers participate in the system. For decentralized and distributed markets, the stability issue needs to be considered when designing the market rules and pricing mechanisms. It is also interesting to investigate how autonomous prosumers will be grouped for a certain goal when they are free to form P2P energy trading coalitions.

On the other hand, various types of advanced methods are there that are suitable for the implementation of energy management in P2P energy transactions, depending on a comprehensive set of factors, such as market characteristics, objectives, and user requirements. With the development of integrated energy markets, more attention should be paid to mechanisms designed for hybrid energy transactions, although the shared energy in most existing P2P energy markets is limited to electrical energy. The distributed optimization algorithm will be extended from a single electricity transaction mode to a hybrid electrical and thermal energy transaction mode [92]. A coordinated multitimescale optimization framework needs to be developed for short-term scheduling and real-time dispatch to achieve operational robustness against power uncertainties [93]. In addition, machine learning techniques can provide opportunities to detect energy consumption patterns and select high-quality participants for demand response programs. It is also worth mentioning that multiagent RL algorithms improve the intelligence and learning capability of market operators and prosumers with DERs. Worldwide zero-carbon transition [94] and smart communities [95] require peer-to-peer energy sharing and trading.

References

[1] Li Y, Qian F, Gao W, Fukuda H, Wang Y. Techno-economic performance of battery energy storage system in an energy sharing community. Journal of Energy Storage 2022;50:104247. Available from: https://doi.org/10.1016/j.est.2022.104247.

[2] Zheng S, Huang G, Lai ACK. Techno-economic performance analysis of synergistic energy sharing strategies for grid-connected prosumers with distributed battery storages. Renewable Energy 2021;178:1261–78. Available from: https://doi.org/10.1016/j.renene.2021.06.100.

[3] Kong KGH, Lim JY, Leong WD, Ng WPQ, Teng SY, Sunarso J, et al. Fuzzy optimization for peer-to-peer (P2P) multi-period renewable energy trading planning. Journal of Cleaner Production 2022;368:133122. Available from: https://doi.org/10.1016/j.jclepro.2022.133122.

[4] Tushar W, Yuen C, Saha TK, Morstyn T, Chapman AC, Alam MJE, et al. Peer-to-peer energy systems for connected communities: a review of recent advances and emerging challenges. Applied Energy 2021;282:116131. Available from: https://doi.org/10.1016/j.apenergy.2020.116131.

[5] Zhang C, Wu J, Zhou Y, Cheng M, Long C. Peer-to-peer energy trading in a microgrid. Applied Energy 2018;220:1–12. Available from: https://doi.org/10.1016/j.apenergy.2018.03.010.

[6] Javed H, Irfan M, Shehzad M, Abdul Muqeet H, Akhter J, Dagar V, et al. Recent trends, challenges, and future aspects of P2P energy trading platforms in electrical-based networks considering blockchain technology: a roadmap toward environmental sustainability. Frontiers in Energy Research 2022;10.

[7] Englberger S, Chapman AC, Tushar W, Almomani T, Snow S, Witzmann R, et al. Evaluating the interdependency between peer-to-peer networks and energy storages: a techno-economic proof for prosumers. Advances in Applied Energy 2021;3:100059. Available from: https://doi.org/10.1016/j.adapen.2021.100059.

[8] Zhou Y. Incentivising multi-stakeholders' proactivity and market vitality for spatiotemporal microgrids in Guangzhou-Shenzhen-Hong Kong Bay Area. Applied Energy 2022;328:120196. Available from: https://doi.org/10.1016/j.apenergy.2022.120196.

[9] Luo C, Zhou X, Lev B. Core, shapley value, nucleolus and nash bargaining solution: a Survey of recent developments and applications in operations management. Omega 2022;110:102638. Available from: https://doi.org/10.1016/j.omega.2022.102638.

[10] Tushar W., Saha T.K., Yuen C., Liddell P., Bean R., Poor H.V. Peer-to-peer energy trading with sustainable user participation: a game theoretic approach. 2018. Available from: https://doi.org/10.48550/arXiv.1809.11034.

[11] Han L, Morstyn T, McCulloch M. Incentivizing prosumer coalitions with energy management using cooperative game theory. IEEE Transactions on Power Systems 2019;34:303–13. Available from: https://doi.org/10.1109/TPWRS.2018.2858540.

[12] Azim MI, Tushar W, Saha TK. Cooperative negawatt P2P energy trading for low-voltage distribution networks. Applied Energy 2021;299:117300. Available from: https://doi.org/10.1016/j.apenergy.2021.117300.

[13] Yang Y, Hu G, Spanos CJ. Optimal sharing and fair cost allocation of community energy storage. IEEE Transactions on Smart Grid 2021;12:4185−94. Available from: https://doi.org/10.1109/TSG.2021.3083882.

[14] Heendeniya CB. Agent-based modeling of a rule-based community energy sharing concept. E3S Web Conf 2021;239:00001. Available from: https://doi.org/10.1051/e3sconf/202123900001.

[15] Zheng S, Huang G, Lai ACK. Coordinated energy management for commercial prosumers integrated with distributed stationary storages and EV fleets. Energy and Buildings 2023;282:112773. Available from: https://doi.org/10.1016/j.enbuild.2023.112773.

[16] Perger T, Wachter L, Fleischhacker A, Auer H. PV sharing in local communities: peer-to-peer trading under consideration of the prosumers' willingness-to-pay. Sustainable Cities and Society 2021;66:102634. Available from: https://doi.org/10.1016/j.scs.2020.102634.

[17] Cosic A, Stadler M, Mansoor M, Zellinger M. Mixed-integer linear programming based optimization strategies for renewable energy communities. Energy 2021;237:121559. Available from: https://doi.org/10.1016/j.energy.2021.121559.

[18] Fan C, Huang G, Sun Y. A collaborative control optimization of grid-connected net zero energy buildings for performance improvements at building group level. Energy 2018;164:536−49. Available from: https://doi.org/10.1016/j.energy.2018.09.018.

[19] Zou L, Munir MdS, Kim K, Hong CS. Day-ahead Energy Sharing Schedule for the P2P Prosumer Community Using LSTM and Swarm Intelligence. 2020 International Conference on Information Networking (ICOIN), 2020, p. 396−401. https://doi.org/10.1109/ICOIN48656.2020.9016520.

[20] Two-stage stochastic programming formulation for optimal design and operation of multi-microgrid system using data-based modeling of renewable energy sources | Elsevier Enhanced Reader n.d. https://doi.org/10.1016/j.apenergy.2021.116830.

[21] Jian P, Guo T, Wang D, Valipour E, Nojavan S. Risk-based energy management of industrial buildings in smart cities and peer-to-peer electricity trading using second-order stochastic dominance procedure. Sustainable Cities and Society 2022;77:103550. Available from: https://doi.org/10.1016/j.scs.2021.103550.

[22] Baharlouei Z, Hashemi M. Efficiency-fairness trade-off in privacy-preserving autonomous demand side management. IEEE Transactions on Smart Grid 2014;5:799−808. Available from: https://doi.org/10.1109/TSG.2013.2296714.

[23] Mulder M. Information asymmetry in retail energy markets. In: Mulder M, editor. Regulation of energy markets: economic mechanisms and policy evaluation. Cham: Springer International Publishing; 2023, p. 125−43. Available from: https://doi.org/10.1007/978-3-031-16571-9_5.

[24] Zhou Y, Wu J, Long C, Ming W. State-of-the-art analysis and perspectives for peer-to-peer energy trading. Engineering 2020;6:739−53. Available from: https://doi.org/10.1016/j.eng.2020.06.002.

[25] Morstyn T, Teytelboym A, Mcculloch MD. Bilateral contract networks for peer-to-peer energy trading. IEEE Transactions on Smart Grid 2019;10:2026−35. Available from: https://doi.org/10.1109/TSG.2017.2786668.

[26] Wang Y, Li Y, Zhao J, Wang G, Jiao W, Qiang Y, et al. A Fast and Secured Peer-to-Peer Energy Trading Using Blockchain Consensus. 2022 IEEE Industry Applications Society Annual Meeting (IAS), 2022, p. 1−8. https://doi.org/10.1109/IAS54023.2022.9939776.

[27] Coelho V, Weiss M, Coelho I, Liu N, Guimarães F. Multi-agent systems applied for energy systems integration: state-of-the-art applications and trends in microgrids. Applied Energy 2017;187:820−32. Available from: https://doi.org/10.1016/j.apenergy.2016.10.056.

[28] Kofinas P, Dounis AI, Vouros GA. Fuzzy Q-learning for multi-agent decentralized energy management in microgrids. Applied Energy 2018;219:53−67. Available from: https://doi.org/10.1016/j.apenergy.2018.03.017.

[29] Münsing E, Mather J, Moura S. Blockchains for decentralized optimization of energy resources in microgrid networks. In: 2017 IEEE conference on control technology and applications (CCTA), 2017, p. 2164−71. https://doi.org/10.1109/CCTA.2017.8062773.

[30] Mihaylov M, Jurado S, Avellana N, Van Moffaert K, de Abril IM, Nowé A. NRGcoin: Virtual currency for trading of renewable energy in smart grids. 11th International Conference on the European Energy Market (EEM14), 2014, p. 1−6. https://doi.org/10.1109/EEM.2014.6861213.

[31] Li Z, Kang J, Yu R, Ye D, Deng Q, Zhang Y. Consortium blockchain for secure energy trading in industrial internet of things. IEEE Transactions on Industrial Informatics 2018;14:3690−700. Available from: https://doi.org/10.1109/TII.2017.2786307.

[32] Karumba S, Sethuvenkatraman S, Dedeoglu V, Jurdak R, Kanhere SS. Barriers to blockchain-based decentralised energy trading: a systematic review. International Journal of Sustainable Energy 2023;42:41−71. Available from: https://doi.org/10.1080/14786451.2023.2171417.

[33] Leong CH, Gu C, Li F. Auction mechanism for P2P local energy trading considering physical constraints. Energy Procedia 2019;158:6613−18. Available from: https://doi.org/10.1016/j.egypro.2019.01.045.

[34] Guerrero J, Chapman AC, Verbič G. Decentralized P2P energy trading under network constraints in a low-voltage network. IEEE Transactions on Smart Grid 2019;10:5163−73. Available from: https://doi.org/10.1109/TSG.2018.2878445.

[35] Long C, Wu J, Zhou Y, Jenkins N. Peer-to-peer energy sharing through a two-stage aggregated battery control in a community microgrid. Applied Energy 2018;226:261−76. Available from: https://doi.org/10.1016/j.apenergy.2018.05.097.

[36] Nguyen S, Peng W, Sokolowski P, Alahakoon D, Yu X. Optimizing rooftop photovoltaic distributed generation with battery storage for peer-to-peer energy trading. Applied Energy 2018;228:2567−80. Available from: https://doi.org/10.1016/j.apenergy.2018.07.042.

[37] Lüth A, Zepter JM, Crespo del Granado P, Egging R. Local electricity market designs for peer-to-peer trading: the role of battery flexibility. Applied Energy 2018;229:1233−43. Available from: https://doi.org/10.1016/j.apenergy.2018.08.004.

[38] Sorin E, Bobo L, Pinson P. Consensus-based approach to peer-to-peer electricity markets with product differentiation. IEEE Transactions on Power Systems 2019;34:994−1004. Available from: https://doi.org/10.1109/TPWRS.2018.2872880.

[39] Esmat A, de Vos M, Ghiassi-Farrokhfal Y, Palensky P, Epema D. A novel decentralized platform for peer-to-peer energy trading market with blockchain technology. Applied Energy 2021;282:116123. Available from: https://doi.org/10.1016/j.apenergy.2020.116123.

[40] Zhou Y, Wu J, Long C. Evaluation of peer-to-peer energy sharing mechanisms based on a multiagent simulation framework. Applied Energy 2018;222:993−1022. Available from: https://doi.org/10.1016/j.apenergy.2018.02.089.

[41] Le Cadre H, Jacquot P, Wan C, Alasseur C. Peer-to-peer electricity market analysis: from variational to Generalized Nash Equilibrium. European Journal of Operational Research 2020;282:753−71. Available from: https://doi.org/10.1016/j.ejor.2019.09.035.

[42] Paudel A, Chaudhari K, Long C, Gooi HB. Peer-to-peer energy trading in a prosumer-based community microgrid: a game-theoretic model. IEEE Transactions on Industrial Electronics 2019;66:6087−97. Available from: https://doi.org/10.1109/TIE.2018.2874578.

[43] Zhu T, Huang Z, Sharma A, Su J, Irwin D, Mishra A, et al. Sharing renewable energy in smart microgrids. In: 2013 ACM/IEEE international conference on cyber-physical systems (ICCPS), 2013, p. 219−28.

[44] Liu N, Yu X, Wang C, Wang J. Energy sharing management for microgrids with PV prosumers: a stackelberg game approach. IEEE Transactions on Industrial Informatics 2017;13:1088−98. Available from: https://doi.org/10.1109/TII.2017.2654302.

[45] Xu X, Li J, Xu Y, Xu Z, Lai CS. A two-stage game-theoretic method for residential PV panels planning considering energy sharing mechanism. IEEE Transactions on Power Systems 2020;35:3562−73. Available from: https://doi.org/10.1109/TPWRS.2020.2985765.

[46] Ahmad T, Zhang H, Yan B. A review on renewable energy and electricity requirement forecasting models for smart grid and buildings. Sustainable Cities and Society 2020;55:102052. Available from: https://doi.org/10.1016/j.scs.2020.102052.

[47] Heydarian-Forushani E, Golshan MEH, Moghaddam MP, Shafie-khah M, Catalão JPS. Robust scheduling of variable wind generation by coordination of bulk energy storages and demand response. Energy Conversion and Management 2015;106: 941−50. Available from: https://doi.org/10.1016/j.enconman.2015.09.074.

[48] Jafari-Marandi R, Hu M, Omitaomu OA. A distributed decision framework for building clusters with different heterogeneity settings. Applied Energy 2016;165:393−404. Available from: https://doi.org/10.1016/j.apenergy.2015.12.088.

[49] Mehrjerdi H. Peer-to-peer home energy management incorporating hydrogen storage system and solar generating units. Renewable Energy 2020;156:183−92. Available from: https://doi.org/10.1016/j.renene.2020.04.090.

[50] Zhou L, Zhou Y. Study on thermo-electric-hydrogen conversion mechanisms and synergistic operation on hydrogen fuel cell and electrochemical battery in energy flexible buildings. Energy Conversion and Management 2023;277:116610. Available from: https://doi.org/10.1016/j.enconman.2022.116610.

[51] Xiao Y, Wang X, Pinson P, Wang X. A local energy market for electricity and hydrogen. IEEE Transactions on Power Systems 2018;33:3898−908. Available from: https://doi.org/10.1109/TPWRS.2017.2779540.

[52] Zhu D, Yang B, Liu Q, Ma K, Zhu S, Ma C, et al. Energy trading in microgrids for synergies among electricity, hydrogen and heat networks. Applied Energy 2020;272:115225. Available from: https://doi.org/10.1016/j.apenergy.2020.115225.

[53] Zhou Y. Sustainable energy sharing districts with electrochemical battery degradation in design, planning, operation and multi-objective optimisation. Renewable Energy 2023;202:1324−41. Available from: https://doi.org/10.1016/j.renene.2022.12.026.

[54] Rodrigues DL, Ye X, Xia X, Zhu B. Battery energy storage sizing optimisation for different ownership structures in a peer-to-peer energy sharing community. Applied Energy 2020;262:114498. Available from: https://doi.org/10.1016/j.apenergy.2020.114498.

[55] Zhou Y. A dynamic self-learning grid-responsive strategy for battery sharing economy—multi-objective optimisation and posteriori multi-criteria decision making. Energy 2023;266:126397. Available from: https://doi.org/10.1016/j.energy.2022.126397.

[56] Barone G, Buonomano A, Forzano C, Giuzio GF, Palombo A. Increasing self-consumption of renewable energy through the Building to Vehicle to Building approach applied to multiple users connected in a virtual micro-grid. Renewable Energy 2020;159:1165−76. Available from: https://doi.org/10.1016/j.renene.2020.05.101.

[57] Transition towards carbon-neutral districts based on storage techniques and spatio-temporal energy sharing with electrification and hydrogenation − ScienceDirect n.d. https://www.sciencedirect.com/science/article/pii/S1364032122003501?casa_token = SiRtKfuZNaAAAAAA:1rr7g5Ck8S6oKjO26RLchJWWEQ-Wjvu9xUhDeo9D PEuV0HulzXWZJhiqjitD1L2itkcXRXiKag (accessed April 26, 2023).

[58] Zhou Y, Zheng S. Machine-learning based hybrid demand-side controller for high-rise office buildings with high energy flexibilities. Applied Energy 2020;262:114416. Available from: https://doi.org/10.1016/j.apenergy.2019.114416.

[59] Razzaq S, Zafar R, Khan NA, Butt AR, Mahmood A. A novel prosumer-based energy sharing and management (PESM) approach for cooperative demand side management (DSM) in smart grid. Applied Sciences 2016;6:275. Available from: https://doi.org/10.3390/app6100275.

[60] Zheng S, Jin X, Huang G, Lai ACK. Coordination of commercial prosumers with distributed demand-side flexibility in energy sharing and management system. Energy 2022;248:123634. Available from: https://doi.org/10.1016/j.energy.2022.123634.

[61] Sánchez Ramos J, Pavón Moreno Mc, Romero Rodríguez L, Guerrero Delgado Mc, Álvarez Domínguez S. Potential for exploiting the synergies between buildings through DSM approaches. Case study: La Graciosa Island. Energy Conversion and Management 2019;194:199−216. Available from: https://doi.org/10.1016/j.enconman.2019.04.084.

[62] Liu N, Yu X, Wang C, Li C, Ma L, Lei J. Energy-sharing model with price-based demand response for microgrids of peer-to-peer prosumers. IEEE Transactions on Power Systems 2017;32:3569−83. Available from: https://doi.org/10.1109/TPWRS.2017.2649558.

[63] Bogensperger A., Ferstl J., Yu Y. Comparison of pricing mechanisms in peer-to-peer energy communities. In: 12th Internationale Energiewirtschaftstagung (IEWT) 2021.

[64] Tushar W, Saha TK, Yuen C, Morstyn T, McCulloch MD, Poor HV, et al. A motivational game-theoretic approach for peer-to-peer energy trading in the smart grid. Applied Energy 2019;243:10−20. Available from: https://doi.org/10.1016/j.apenergy.2019.03.111.

[65] O'Mahoney A, Denny E. Electricity prices and generator behaviour in gross pool electricity markets. Energy Policy 2013;63:628−37. Available from: https://doi.org/10.1016/j.enpol.2013.08.098.

[66] Muhsen H, Allahham A, Al-Halhouli A, Al-Mahmodi M, Alkhraibat A, Hamdan M. Business model of peer-to-peer energy trading: a review of literature. Sustainability 2022;14:1616. Available from: https://doi.org/10.3390/su14031616.

[67] Guerrero J, Chapman AC, Verbic G. Trading arrangements and cost allocation in P2P energy markets on low-voltage networks. In: 2019 IEEE Power & Energy Society General Meeting (PESGM), 2019, p. 1−5. https://doi.org/10.1109/PESGM40551.2019.8973410.

[68] Zhang C, Yang T, Wang Y. Peer-to-peer energy trading in a microgrid based on iterative double auction and blockchain. Sustainable Energy, Grids and Networks 2021;27:100524. Available from: https://doi.org/10.1016/j.segan.2021.100524.

[69] Navon A, Ben Yosef G, Machlev R, Shapira S, Roy Chowdhury N, Belikov J, et al. Applications of game theory to design and operation of modern power systems: a comprehensive review. Energies 2020;13:3982. Available from: https://doi.org/10.3390/en13153982.

[70] Zhang C, Wu J, Cheng M, Zhou Y, Long C. A bidding system for peer-to-peer energy trading in a grid-connected microgrid. Energy Procedia 2016;103:147−52. Available from: https://doi.org/10.1016/j.egypro.2016.11.264.

[71] Opadokun F, Roy TK, Akter MN, Mahmud MA. Prioritizing customers for neighborhood energy sharing in residential microgrids with a transactive energy market. In: 2017 IEEE power & energy society general meeting, 2017, p. 1−5. https://doi.org/10.1109/PESGM.2017.8274582.

[72] Lin J, Pipattanasomporn M, Rahman S. Comparative analysis of auction mechanisms and bidding strategies for P2P solar transactive energy markets. Applied Energy 2019;255:113687. Available from: https://doi.org/10.1016/j.apenergy.2019.113687.

[73] Karlsson M, Ygge F, Andersson A. Market-based approaches to optimization. Computational Intelligence 2007;23:92−109. Available from: https://doi.org/10.1111/j.1467-8640.2007.00296.x.

[74] Dagdougui H, Sacile R. Decentralized control of the power flows in a network of smart microgrids modeled as a team of cooperative agents. IEEE Transactions on Control Systems Technology 2014;22:510−19. Available from: https://doi.org/10.1109/TCST.2013.2261071.

[75] Baroche T, Pinson P, Latimier RLG, Ahmed HB. Exogenous cost allocation in peer-to-peer electricity markets. IEEE Transactions on Power Systems 2019;34:2553−64. Available from: https://doi.org/10.1109/TPWRS.2019.2896654.

[76] Pourbabak H, Luo J, Chen T, Su W. A novel consensus-based distributed algorithm for economic dispatch based on local estimation of power mismatch. IEEE Transactions on Smart Grid 2018;9:5930−42. Available from: https://doi.org/10.1109/TSG.2017.2699084.

[77] Vinyals M., Velay M., Sisinni M. A multi-agent system for energy trading between prosumers. In: Omatu S, Rodríguez S, Villarrubia G, Faria P, Sitek P, Prieto J, editors. Distributed computing and artificial intelligence, 14th international conference. Cham: Springer International Publishing; 2018, p. 79−86. Available from: https://doi.org/10.1007/978-3-319-62410-5_10.

[78] Celik B., Roche R., Bouquain D., Miraoui A. Coordinated neighborhood energy sharing using game theory and multi-agent systems. In: 2017 IEEE manchester PowerTech, 2017, p. 1−6. Available from: https://doi.org/10.1109/PTC.2017.7980820.

[79] Qiu D, Ye Y, Papadaskalopoulos D, Strbac G. Scalable coordinated management of peer-to-peer energy trading: a multi-cluster deep reinforcement learning approach. Applied Energy 2021;292:116940. Available from: https://doi.org/10.1016/j.apenergy.2021.116940.

[80] Zhou Y, Lund PD. Peer-to-peer energy sharing and trading of renewable energy in smart communities − trading pricing models, decision-making and agent-based collaboration. Renewable Energy 2023;207:177−93. Available from: https://doi.org/10.1016/j.renene.2023.02.125.

[81] Jiang S., Venticinque S., Horn G., Hallsteinsen S., Noebels M. A distributed agent-based system for coordinating smart solar-powered microgrids. In: 2016 SAI Computing Conference (SAI), 2016, p. 71−9. Available from: https://doi.org/10.1109/SAI.2016.7555964.

[82] Cintuglu MH, Martin H, Mohammed OA. Real-time implementation of multiagent-based game theory reverse auction model for microgrid market operation. IEEE Transactions on Smart Grid 2015;6:1064−72. Available from: https://doi.org/10.1109/TSG.2014.2387215.

[83] Zhou Y. Advances of machine learning in multi-energy district communities-mechanisms, applications and perspectives. Energy and AI 2022;10:100187. Available from: https://doi.org/10.1016/j.egyai.2022.100187.

[84] Todd-Blick A, Spurlock CA, Jin L, Cappers P, Borgeson S, Fredman D, et al. Winners are not keepers: characterizing household engagement, gains, and energy patterns in demand response using machine learning in the United States. Energy Research & Social Science 2020;70:101595. Available from: https://doi.org/10.1016/j.erss.2020.101595.

[85] Luo F, Ranzi G, Wang X, Dong ZY. Social information filtering-based electricity retail plan recommender system for smart grid end users. IEEE Transactions on Smart Grid 2019;10:95−104. Available from: https://doi.org/10.1109/TSG.2017. 2732346.

[86] Chen T, Su W. Indirect customer-to-customer energy trading with reinforcement learning. IEEE Transactions on Smart Grid 2019;10:4338−48. Available from: https://doi.org/10.1109/TSG.2018.2857449.

[87] Charbonnier F, Morstyn T, McCulloch MD. Scalable multi-agent reinforcement learning for distributed control of residential energy flexibility. Applied Energy 2022;314:118825. Available from: https://doi.org/10.1016/j.apenergy.2022.118825.

[88] Jogunola O, Adebisi B, Ikpehai A, Popoola SI, Gui G, Gačanin H, et al. Consensus algorithms and deep reinforcement learning in energy market: a review. IEEE Internet of Things Journal 2021;8:4211−27. Available from: https://doi.org/10. 1109/JIOT.2020.3032162.

[89] Morstyn T, Savelli I, Hepburn C. Multiscale design for system-wide peer-to-peer energy trading. One Earth 2021;4:629−38. Available from: https://doi.org/10.1016/ j.oneear.2021.04.018.

[90] Chen K, Lin J, Song Y. Trading strategy optimization for a prosumer in continuous double auction-based peer-to-peer market: a prediction-integration model. Applied Energy 2019;242:1121−33. Available from: https://doi.org/10.1016/j.apenergy. 2019.03.094.

[91] Chen Y, Yang Y, Xu X. Towards transactive energy: an analysis of information-related practical issues. Energy Conversion and Economics 2022;3:112−21. Available from: https://doi.org/10.1049/enc2.12057.

[92] Liu N, Wang J, Wang L. Hybrid energy sharing for multiple microgrids in an integrated heat−electricity energy system. IEEE Transactions on Sustainable Energy 2019;10:1139−51. Available from: https://doi.org/10.1109/TSTE.2018.2861986.

[93] Bao Z, Ye Y, Wu L. Multi-timescale coordinated schedule of interdependent electricity-natural gas systems considering electricity grid steady-state and gas network dynamics. International Journal of Electrical Power & Energy Systems 2020;118: 105763. Available from: https://doi.org/10.1016/j.ijepes.2019.105763.

[94] Zhou Y. Worldwide carbon neutrality transition? Energy efficiency, renewable, carbon trading and advanced energy policies. Energy Reviews 2023;100026.

[95] Zhou Y. Low-carbon transition in smart city with sustainable airport energy ecosystems and hydrogen-based renewable-grid-storage-flexibility. Energy Reviews 2022; 100001.

CHAPTER 9

Blockchain technologies for automatic, secure, and tamper-proof energy trading

Lu Zhou[1] and Yuekuan Zhou[1,2,3,4]
[1]Sustainable Energy and Environment Thrust, Function Hub, The Hong Kong University of Science and Technology (Guangzhou), Nansha, Guangdong, P.R. China
[2]Department of Mechanical and Aerospace Engineering, The Hong Kong University of Science and Technology, Clear Water Bay, Hong Kong SAR, P.R. China
[3]HKUST Shenzhen-Hong Kong Collaborative Innovation Research Institute, Futian, Shenzhen, P.R. China
[4]Division of Emerging Interdisciplinary Areas, The Hong Kong University of Science and Technology, Clear Water Bay, Hong Kong SAR, P.R. China

Nomenclature

Abbreviations
B2G building-to-grid
B2V building-to-vehicle
G2B grid-to-building
P2P peer-to-peer
V2B vehicle-to-building

9.1 Introduction

The energy decarbonization scenario is identified as having a higher diversity of energy supplies and lower energy trade [1]. Peer-to-peer (P2P) energy sharing has attracted widespread interest in incentivizing the participation of prosumers [2,3], enhancing renewable penetration [4,5], stabilizing grid power [6], promoting sustainability [7], and so on. Compared with the traditional centralized P2P energy trading approach with requirements on private data of prosumers (e.g., system configuration, operational status, and dynamic profiles on energy supply/consumption), the consciousness of prosumers about data privacy protection will impose great challenges on centralized P2P energy trading. In contrast, distributed P2P energy trading can enable vitality and prosperity in the double auction energy market through energy coordination, complementarity, and privacy preservation. Blockchain can also promote the sustainability transition through innovation management and multistakeholder participation [8].

Advances in Digitalization and Machine Learning for Integrated
Building-Transportation Energy Systems
DOI: https://doi.org/10.1016/B978-0-443-13177-6.00007-2

Generally, blockchain technology can be applied in distributed renewable systems for secure and autonomous energy trading with privacy protection of the players. Petri et al. [9] proposed a blockchain-based energy framework with smart contracts for secure energy trading. For the system to survive malicious attacks, Kavousi-Fard et al. [10] developed a blockchain platform with secured consensus from market participants. The research results demonstrated its fault-tolerant ability against cyberattacks. Wongthongtham et al. [11] comprehensively reviewed blockchain-based P2P energy trading, concerning scalability, security, and decentralization. A proof-of-stake protocol has also been proposed [12] to improve the income of producers and cost savings of consumers. Autonomous P2P energy trading with blockchain has attracted the interest of researchers. To compensate for the traditional centralized power grid, Vieira and Zhang [13] used the concepts of smart contracts and double auctions in autonomous trading. Mehdinejad et al. [14] used the demurrage mechanism to economically incentivize prosumers and customers to participate in demand response and local P2P trading. The private trading data of participants is protected in the energy token market. Yang et al. [15] developed a decentralized energy trading platform for residential communities. Results showed that the platform can reduce an individual's cost by 38.6% and the overall cost by 11.2%. Umar et al. [16] developed a blockchain-based decentralized trading platform for transparent, secure, and fast transactions. The platform can promote a self-sustained microgrid with economic incentives.

In the context of efficiency and privacy issues in blockchain, Dong et al. [17] developed an account mapping algorithm and applied it to decentralized energy trading. The two-layer transaction case validates privacy protection and trading efficiency. Lei et al. [18] developed homomorphic encryption for privacy protection without affecting trading efficiency. Samuel et al. [19] developed a secure energy trading platform with dynamic pricing for privacy protection, operating cost savings for each stakeholder, and social welfare maximization. Results showed that, compared with fixed pricing, dynamic pricing can save more than 16% on operating costs with less information loss. Regarding blockchain in the energy sector, Ahl et al. [20] provided a holistic overview of blockchain-based P2P microgrids and a multidimensional analytical framework.

Furthermore, automatic trading based on smart contracts and Internet of Things (IoT) technology has attracted the interest of researchers. The permitted blockchain-based trading platform [18] can incentivize user participation in automatic trading with safety, reliability, and efficiency. Han et al. [21] developed an automatic trading platform in a decentralized energy market without human interaction. The platform can balance the profits of players and enhance renewable penetration. To balance the mutual benefits and sustain the automatic trading, Alt et al. [22] highlighted the importance of mutual trust between energy generators and consumers and reaching a consensus in the contract. Mengelkamp et al. [23] formulated a blockchain-based microgrid market without central intermediaries. However, the automatic P2P market is highly constrained by the current regulations.

9.2 Mechanisms and configurations of blockchain techniques

A blockchain is a digital ledger technology with many records, or blocks, securely linked together using cryptography. Each block contains a cryptographic hash of the previous block, a timestamp, and transaction data [24–27]. Blockchains are typically managed by a P2P computer network for use as a public distributed ledger, where nodes collectively adhere to a consensus algorithm protocol to add and validate new transaction blocks. In other words, the transaction is successful only with the agreement of all participants.

Blockchain is considered as a promising tool to record and facilitate transactions between generators and consumers of energy, which includes various use cases, such as P2P electricity trading and selling of excessive renewable energy to other network participants through automated smart contracts. Up until now, blockchain has shown numerous advantages:
1. Verification and traceability of multistep transactions
2. Secure transactions
3. Reducing compliance costs
4. Speed up data transfer processing

Fig. 9.1 shows public permissionless ledgers and private permissionless ledgers. Public permissionless ledgers can provide free participation of each user in the network, while private permissionless ledgers require validation for authorized nodes.

Andoni et al. [29] comprehensively reviewed blockchain-based energy systems, including architectures, distributed consensus algorithms, and critical components. Due to the emergence of rooftop photovoltaics, electric vehicles (EVs), hydrogen vehicles, distributed energy storages, and smart

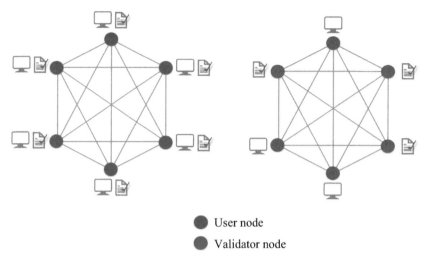

● User node
● Validator node

Figure 9.1 Public permissionless ledgers and private permissioned ledgers [28].

metering, blockchain can play an essential role in future energy markets through smart contracts and system interoperability, with respect to novel business models for energy markets, smart energy management, carbon credits, or renewable energy certificates. By tracking the energy supply chain, distributed ledger technology can improve efficiency for energy suppliers. The most promising aspect of blockchain is the replacement of grid infrastructure ownership and services from retailers to consumers (such as billing and metering usage), enabling the energy trade and purchase with grids directly instead of retailers.

9.3 Data privacy and energy security in peer-to-peer energy trading

Although P2P energy trading shows advantages in renewable penetration, sustainable transition, and cost savings, several critical concerns, such as data privacy, energy trading security, etc., need to be considered. Zhou et al. [30] comprehensively reviewed the market design, trading schemes, physical infrastructure, information, and communication technology of P2P energy trading. Deng et al. [31] developed homomorphic encryption technology to protect the energy purchase price and amount. The privacy protection technique can promote more participants and develop customer-side energy management strategies for cost savings. By reaching the global optimal solution through neighborhood negotiation, Umer et al. [32] developed a new communication-efficient algorithm, enabling the trading scheme to be more privacy-preserving and scalable. Guan et al. [33] developed a privacy-preserving energy trading scheme using ciphertext-policy attribute-based encryption. The scheme can provide access control through transaction arbitration in ciphertext form. Experimental results provided evidence of the validity and practicability of the proposed approach. To protect data privacy during the renewable energy trading process, Deng et al. [31] integrated homomorphic encryption technology to hide the bidding energy purchase price and amount for each individual. Qiu et al. [34] developed a coordinated management scheme for P2P energy trading using multiagent deep reinforcement learning. By adding a trusted third party for market trading information collection, the proposed approach shows cost-saving and time-saving performance with generalization capability.

Fig. 9.2 demonstrates the structural configuration of an energy internet framework [33], consisting of energy suppliers (e.g., wind power, solar power, hydropower, and nuclear power), intermediate energy markets, and energy users (e.g., buildings, industries, and vehicles). In the energy trading market, the intermediate retailer manages the relationship between energy sellers and energy purchasers. During the trading process, both energy sellers and purchasers will have bidirectional power interactions in

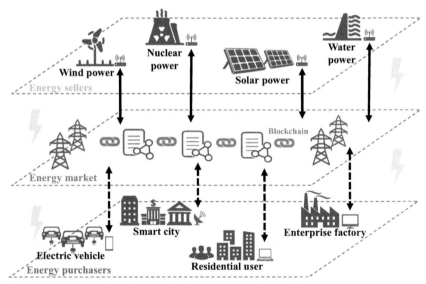

Figure 9.2 Structural configuration of an energy internet [33].

the energy market, leading to potential risks such as data leakage, privacy disclosure, and data tampering.

Researchers studied threat models and energy security in the decentralized energy market.

Samuel and Javaid [35] studied the security of blockchain-based demurrage mechanisms with the consideration of a threat model. Results show a decrease of overhead costs of up to 66.67% with minimal average hash power of up to 82.75%. To ensure security and transparency, Guan et al. [36] proposed a blockchain-based energy trading scheme (BC-ETS) with two levels for privacy protection and supply-demand balance. Results showed the superiority of the proposed BC-ETS over other schemes. The energy revolution with 6G digitalization can improve security, connectivity, integration, and sensory data processing [37], while incompatibility with older devices, power consumption, and operating costs lead to constraints for widespread applications. Kirli et al. [38] reviewed applications of smart contracts in energy systems for 13 industrial projects. Silvestre et al. [39] reviewed blockchain applications in power systems. The advantages of blockchain in energy systems include tamper-proof data, privacy protection, and efficient computation [40].

9.4 Secure and tamper-proof energy trading with blockchain and automated smart contracts

A smart contract is a computer program or a transactional protocol intended to automatically execute, control, or document legally relevant events and actions according to the terms of a contract or an agreement. Fig. 9.3 demonstrates an energy blockchain with a smart contract. The data and events are external inputs. Through the state machine and contract transaction set, the current asset state and the execution selection of the following contract transaction are judged and docked with external information. The smart contract can be integrated with game theory with cryptocurrency payment functions to motivate nodes to write new ledgers. Furthermore, game-theoretic incentives or rewards can be applied to promote collaboration and honest behavior of distributed nodes.

Based on energy blockchain and smart contract, applications for secure and tamper-proof energy trading environments have also been studied. Esmat et al. [42] developed a new decentralized platform consisting of a market and blockchain. The smart contract in blockchain can promote automation, security, and fast real-time settlements. Han et al. [21] designed

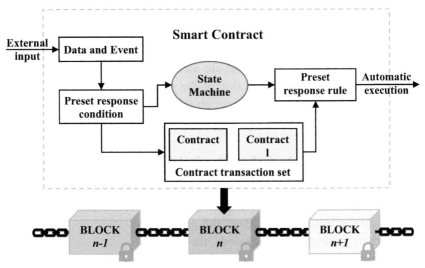

Figure 9.3 Energy blockchain with smart contract [41].

a smart contract architecture for human-free energy trading and payment, ensuring security and fairness in the allocation of economic benefits and promoting renewable energy trading processes. Wu et al. [43] designed a microgrid-blockchain framework for energy trading in electrified transportation, building, and industrial sectors. Wang et al. [44] developed a privacy-preserving decentralized energy trading scheme with market clearing functions. The blockchain shows three functions: (1) limited access to private data; (2) quotation calculation in parallel; and (3) transparent on-chain market clearing. The uniform clearing of decentralized transactions under smart contracts is attractive in a decentralized market. Muzumdar et al. [45] developed a smart grid energy trading framework based on distributed ledgers and smart contracts. The framework can incentivize energy trading while protecting privacy and data transparency. Wu et al. [46] comprehensively reviewed blockchain-enabled P2P energy trading with information and communication technology and multiscale P2P flexibility. To build up interoperative marketplaces, energy digitalization with IoT-based P2P interoperability and blockchain-enabled trading privacy and security are necessary elements for secure and reliable energy markets. Samuel et al. [19] proposed a secure energy system with blockchains for information attack protection and efficient energy trading for EVs and district communities. By applying differential privacy in the EMS algorithm, the energy price and operational cost can be reduced, together with less information loss and higher privacy protection. Tsao and Thanh [47] applied blockchain in P2P energy trading to replace the control from a central regulatory authority with a separate personal control. The intermittency of renewable energy and uncertainty in demand are addressed by a fuzzy multiobjective programming model. The feasibility of the proposed approach and sustainable microgrid design is verified through experiments.

9.5 Prospects of blockchain techniques in multienergy trading platforms

In achieving sustainable development goals, blockchains play a significant role in energy security, reliability, and privacy protection, especially with intermittent renewable integrations through distributed units. Fig. 9.4 demonstrates a small-scale district with universe subcomponents for carbon neutrality transformation. Onsite renewables can be partially self-consumed, and the rest can either charge EVs for daily transportation or be shared with neighborhood buildings through P2P energy sharing. With the mobility of EVs, spatiotemporal energy sharing can be achieved in cross-boundary cities. For example, the abundant energy in lowly densified suburbs (e.g., Guangzhou) can be shared with highly densified city centers (e.g., Hong Kong) for regional or cross-regional energy balance. Furthermore, digitization, information, and intelligence will play a significant role in an electrified network. In an energy-sharing platform with distributed renewable systems, zero-emission vehicles, buildings, and digital currency, several critical challenges need to be considered (Fig. 9.4(B)).
1. Energy equality among different stakeholders
2. Fair economic balance and allocation
3. Security, reliability and privacy
4. Win-win benefits and collaboration
    Challenges and future prospects regarding sustainable energy transformation are summarized as follows:
1. Significant challenges, such as legal, regulatory, and competition barriers, are imposed on blockchain technologies before they are widely applied with commercial and financial viability.
2. Difficulty in fairly balancing economic efficiency and information privacy.
3. New P2P market designs are required in a blockchain-based fraud-resilient and secure market, imposing pressure on the traditional monopolized electricity market.

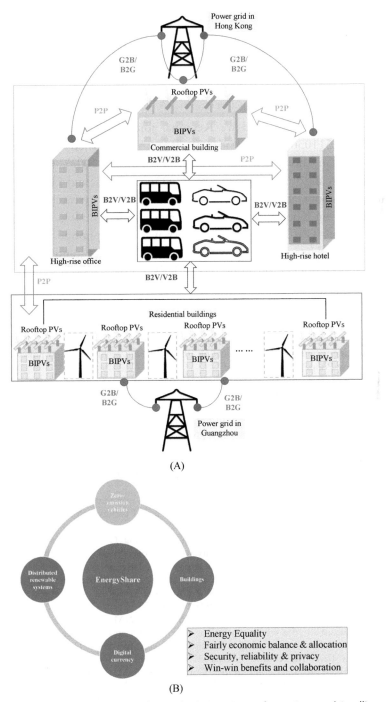

Figure 9.4 District carbon neutrality with digitization, information, and intelligence: (A) electrification network; (B) energy-sharing platform.

4. Both the external environment (e.g., policy support and talent cultivation) and internal technology (e.g., core technology and security control) need to synergistically cooperate to promote blockchain and energy.

Acknowledgment

This work was supported by Regional joint fund youth fund project (2022A1515110364, P00038-1002), Basic and Applied Basic Research Project-Guangzhou 2023 (2023A04J1035, P00121-1003), Joint Funding of Institutes and Enterprises in 2023 (2023A03J0104), and HKUST (GZ)-enterprise cooperation project (R00017-2001). This research is supported by The Hong Kong University of Science and Technology (Guangzhou) startup grant (G0101000059). This work was also supported in part by the Project of Hetao Shenzhen-Hong Kong Science and Technology Innovation Cooperation Zone (HZQB-KCZYB-2020083).

References

[1] Jewell J, Cherp A, Riahi K. Energy security under de-carbonization scenarios: an assessment framework and evaluation under different technology and policy choices. Energy Policy 2014;65:743−60.
[2] Morstyn T, Farrell N, Darby SJ, et al. Using peer-to-peer energy-trading platforms to incentivize prosumers to form federated power plants. Nature Energy 2018;3:94−101.
[3] Li G, Li Q, Song W. L Wang. Incentivizing distributed energy trading among prosumers: a general Nash bargaining approach. International Journal of Electrical Power & Energy Systems 2021;131:107100.
[4] Liu J, Yang H, Zhou Y. Peer-to-peer energy trading of net-zero energy communities with renewable energy systems integrating hydrogen vehicle storage. Applied Energy 2021;298:117206.
[5] Liu J, Yang H, Zhou Y. Peer-to-peer trading optimizations on net-zero energy communities with energy storage of hydrogen and battery vehicles. Applied Energy 2021;302:117578.
[6] He Y, Zhou Y, Yuan J, Liu Z, Wang Z, Zhang G. Transformation towards a carbon-neutral residential community with hydrogen economy and advanced energy management strategies. Energy Conversion and Management 2021;249:114834.
[7] Kim H, Choi H, Kang H, An J, Yeom S, Hong T. A systematic review of the smart energy conservation system: from smart homes to sustainable smart cities. Renewable and Sustainable Energy Reviews 2021;140:110755.
[8] Ahl A, Yarime M, Goto M, Chopra SS, Kumar NM, Tanaka K, et al. Exploring blockchain for the energy transition: opportunities and challenges based on a case study in Japan. Renewable and Sustainable Energy Reviews 2020;117:109488.
[9] Petri I, Barati M, Rezgui Y, Rana OF. Blockchain for energy sharing and trading in distributed prosumer communities. Computers in Industry 2020;123:103282.
[10] Kavousi-Fard A, Almutairi A, Al-Sumaiti A, Faroughian A, Alyami S. An effective secured peer-to-peer energy market based on blockchain architecture for the

interconnected microgrid and smart grid. International Journal of Electrical Power & Energy Systems 2021;132:107171.

[11] Wongthongtham P, Marrable D, Abu-Salih B, Liu X, Morrison G. Blockchain-enabled peer-to-peer energy trading. Computers & Electrical Engineering 2021;94:107299.

[12] Yang J, Paudel A, Gooi HB, Nguyen HD. A proof-of-stake public blockchain based pricing scheme for peer-to-peer energy trading. Applied Energy 2021;298:117154.

[13] Vieira G, Zhang J. Peer-to-peer energy trading in a microgrid leveraged by smart contracts. Renewable and Sustainable Energy Reviews 2021;143:110900.

[14] Mehdinejad M, Shayanfar H, Mohammadi-Ivatloo B. Decentralized blockchain-based peer-to-peer energy-backed token trading for active prosumers. Energy 2022;244:122713.

[15] Yang Q, Wang H, Wang T, Zhang S, Wu X, Wang H. Blockchain-based decentralized energy management platform for residential distributed energy resources in a virtual power plant. Applied Energy 2021;294:117026.

[16] Umar A, Kumar D, Ghose T. Blockchain-based decentralized energy intra-trading with battery storage flexibility in a community microgrid system. Applied Energy 2022;322:119544.

[17] Dong J, Song C, Zhang T, Hu Y, Zheng H, Li Y. Efficient and privacy-preserving decentralized energy trading scheme in a blockchain environment. Energy Reports 2022;8:485−93.

[18] Lei YT, Ma CQ, Mirza N, Ren YS, Narayan SW, Chen XQ. A renewable energy microgrids trading management platform based on permissioned blockchain. Energy Economics 2022;115:106375.

[19] Samuel O, Javaid N, Almogren A, Javed MU, Qasim U, Radwan A. A secure energy trading system for electric vehicles in smart communities using blockchain. Sustainable Cities and Society 2022;79:103678.

[20] Ahl A, Yarime M, Tanaka K, Sagawa D. Review of blockchain-based distributed energy: Implications for institutional development. Renewable and Sustainable Energy Reviews 2019;107:200−11.

[21] Han D, Zhang C, Ping J, Yan Z. Smart contract architecture for decentralized energy trading and management based on blockchains. Energy 2020;199:117417.

[22] Alt R, Wende E. Blockchain technology in energy markets − an interview with the European Energy Exchange. Electron Markets 2020;30:325−30.

[23] Mengelkamp E, Gärttner J, Rock K, Kessler S, Orsini L, Weinhardt C. Designing microgrid energy markets: a case study: the Brooklyn microgrid. Applied Energy 2018;210:870−80.

[24] Morris D.Z.. Leaderless, Blockchain-Based Venture Capital Fund Raises $100 Million, And Counting. Fortune;21 May 2016. Retrieved 23 May 2016.

[25] Popper N.. A venture fund with plenty of virtual capital, but no capitalist. The New York Times; 22 May 2016. Retrieved 23 May 2016.

[26] Vranken H. Sustainability of bitcoin and blockchains. Current Opinion in Environmental Sustainability 2017;28:1−9.

[27] Narayanan A, Bonneau J, Felten E, Miller A, Goldfeder S. Bitcoin and cryptocurrency technologies: a comprehensive introduction. Princeton, NJ: Princeton University Press; 2016.

[28] Froystad P., Holm J. Blockchain: powering the internet of value (White paper).

[29] Andoni M, Robu V, Flynn D, Abram S, Geach D, Jenkins D, et al. Blockchain technology in the energy sector: a systematic review of challenges and opportunities. Renewable and Sustainable Energy Reviews 2019;100:143−74.

[30] Zhou Y, Wu J, Long C, Ming W. State-of-the-art analysis and perspectives for peer-to-peer energy trading. Engineering 2020;6(7):739−53.

[31] Deng R, Luo F, Yang J, Huang DW, Ranzi G, Dong ZY. Privacy preserving renewable energy trading system for residential communities. International Journal of Electrical Power & Energy Systems 2022;142:108367.

[32] Umer K, Huang Q, Khorasany M, Afzal M, Amin W. A novel communication efficient peer-to-peer energy trading scheme for enhanced privacy in microgrids. Applied Energy 2021;296:117075.

[33] Guan Z, Lu X, Yang W, Wu L, Wang N, Zhang Z. Achieving efficient and Privacy-preserving energy trading based on blockchain and ABE in smart grid. Journal of Parallel and Distributed Computing 2021;147:34−45.

[34] Qiu D, Ye Y, Papadaskalopoulos D, Strbac G. Scalable coordinated management of peer-to-peer energy trading: a multi-cluster deep reinforcement learning approach. Applied Energy 2021;292:116940.

[35] Samuel O, Javaid N. A secure blockchain-based demurrage mechanism for energy trading in smart communities. International Journal of Energy Research 2021;45 (1):297−315.

[36] Guan Z, Lu X, Wang N, Wu J, Du X, Guizani M. Towards secure and efficient energy trading in IIoT-enabled energy internet: a blockchain approach. Future Generation Computer Systems 2020;110:686−95.

[37] Yap KY, Chin HH, Klemeš JJ. Future outlook on 6G technology for renewable energy sources (RES). Renewable and Sustainable Energy Reviews 2022;167:112722.

[38] Kirli D, Couraud B, Robu V, Salgado-Bravo M, et al. Smart contracts in energy systems: a systematic review of fundamental approaches and implementations. Renewable and Sustainable Energy Reviews 2022;158:112013.

[39] Silvestre MLD, Gallo P, Guerrero JM, Musca R, Sanseverino ER, Sciumè G, et al. Blockchain for power systems: current trends and future applications. Renewable and Sustainable Energy Reviews 2020;119:109585.

[40] Dinesha DL, Balachandra P. Conceptualization of blockchain enabled interconnected smart microgrids. Renewable and Sustainable Energy Reviews 2022;168:112848.

[41] Chen X, Zhang X. Secure electricity trading and incentive contract model for electric vehicle based on energy blockchain. IEEE Access 2019;7:178763−78.

[42] Esmat A, de Vos M, Ghiassi-Farrokhfal Y, Palensky P, Epema D. A novel decentralized platform for peer-to-peer energy trading market with blockchain technology. Applied Energy 2021;282:116123.

[43] Wu Y, Wu Y, Cimen H, Vasquez JC, Guerrero JM. Towards collective energy community: potential roles of microgrid and blockchain to go beyond P2P energy trading. Applied Energy 2022;314:119003.

[44] Wang B, Zhao S, Li Y, Wu C, Tan J, Li H, et al. Design of a privacy-preserving decentralized energy trading scheme in blockchain network environment. International Journal of Electrical Power & Energy Systems 2021;125:106465.

[45] Muzumdar A, Modi C, Madhu GM, Vyjayanthi C. A trustworthy and incentivized smart grid energy trading framework using distributed ledger and smart contracts. Journal of Network and Computer Applications 2021;183−184:103074.

[46] Wu Y, Wu Y, Cimen H, Vasquez JC, Guerrero JM. P2P energy trading: Blockchain-enabled P2P energy society with multi-scale flexibility services. Energy Reports 2022;8:3614−28.

[47] Tsao YC., Thanh VV. Toward sustainable microgrids with blockchain technology-based peer-to-peer energy trading mechanism: A fuzzy meta-heuristic approach. Renewable and Sustainable Energy Reviews 2021(136):110452.

CHAPTER 10

Energy economy and robustness with mobile energy storage systems

Xiaoyuan Xu[1], Tingxuan Chen[1] and Zhuoxin Lu[2]
[1]Department of Electrical Engineering, Shanghai Jiao Tong University, Shanghai, P.R. China
[2]Department of Electrical Engineering, University of Shanghai for Science and Technology, Shanghai, P.R. China

10.1 Introduction

With the progress of high-density and high-energy battery energy storage techniques, the mobile energy storage system (MESS) has attracted more attention. Because of its flexibility and mobility, the MESS can enhance power system economy and resilience through service restoration [1,2], load recovery [3,4], and network congestion management [5,6]. Besides, it reduces the operating costs of power grids by smoothing out the fluctuation of renewable energy power generation [7].

In the existing literature, two-stage stochastic or robust optimization methods are commonly used to solve the MESS scheduling problem, considering the variabilities of photovoltaic (PV) power and load demands. Specifically, the MESS routing strategy is determined at the first stage, and the charging and discharging strategy is optimized at the second stage considering the realization of random variables over the entire time horizon. However, the nonanticipativity constraints on MESS routing and the state of charge (SOC) would render the two-stage scheduling strategy infeasible when facing the randomness of travel time, renewable power, and load demands [8,9]. The multistage robust optimization is generally used to achieve nonanticipativity results but its application to MESS scheduling needs to be highlighted. Also, the role of MESS in promoting PV power integration and enhancing the economic operation of the distribution system has not been extensively discussed.

Although MESS scheduling problems are commonly designed as optimization problems with objective functions and constraints, there exist

Advances in Digitalization and Machine Learning for Integrated Building-Transportation Energy Systems
DOI: https://doi.org/10.1016/B978-0-443-13177-6.00020-5

some limitations in these model-based methods. First, MESS operation problems are usually established based on accurate predictions of electricity prices and travel time, which are difficult to obtain in actual situations. The uncertainties associated with electricity prices and travel time affect MESS operation [10], and their explicit modeling increases the problem solving difficulty. Second, the MESS battery model is usually simplified as a linear model to achieve computational tractability. Both operating efficiency and life degradation of electro-chemical energy storage have nonlinear relations with operating temperature, charging/discharging power, depth of discharge, and SOC [11,12]. Hence a refined battery model should be considered to better estimate the battery life and obtain more realistic operating scenarios. Furthermore, the computational efficiency in solving MESS operation problems cannot be guaranteed due to a large number of integer variables.

Deep reinforcement learning (DRL), which is a research hotspot in decision-making and control, has the potential to solve the above issues. DRL is a branch of machine learning concerned with how agents take actions in an uncertain interactive environment and learn from the feedback to optimize their policies [13,14]. It deals with nonlinear and stochastic problems in a data-driven approach that learns the endogenous characteristics through the historical or constructed dataset. Therefore DRL is a promising technique to solve ESS operation problems with detailed energy storage models. However, to our knowledge, the research on leveraging DRL in MESS arbitrage has not been highlighted. In addition, the impacts of uncertain factors in power and transportation networks and the degradation of battery life on MESS operation also need further exploration.

In this context, this chapter explores the MESS application in the economic operation of PV-penetrated distribution systems. As a shared energy storage system (ESS), MESS travels and shifts excess PV power among charging stations, leveraging the available traffic monitoring information in an urban transportation network. Besides, this chapter also proposes a data-driven uncertainty-adaptive MESS arbitrage method considering complicated battery models. A two-layer DRL method is designed to obtain routing and scheduling strategies for MESS operations. The uncertainties associated with electricity prices and traffic conditions are incorporated into the proposed framework.

10.2 Model-based routing and scheduling of mobile energy storage system in coupled transportation and power distribution networks

10.2.1 Deterministic optimization model

The proposed MESS scheduling method aims to achieve large-scale PV power integration into the distribution network [15]. Different from stationary ESSs, the MESS travels and delivers the excess PV power among different charging stations, leveraging the available traffic monitoring information in an urban transportation network. The assumptions on the MESS operation are outlined as follows:

1. MESS is a battery that is carried on an electric truck; the MESS cannot immediately turn around when traveling on a path.
2. MESS can work in charging, discharging, or idle mode when residing at a charging station.
3. MESS travels using the current average travel speed on a path; the current and historical travel speeds on any path can be downloaded from the online traffic monitoring system.

10.2.1.1 Mobile energy storage system operation constraints

Fig. 10.1 gives a simple example of temporal-spatial MESS mobility in the transportation network. S1, S2, and S3 are MESS charging stations, and S1 is the initial/terminal station as MESS makes daily travels among stations. Here, S1-S1 indicates that MESS stays at S1, where it is operated in charging, discharging, or idle mode. S1-S2 indicates that MESS travels from S1 to S2. In addition, a day is divided into 96 time periods as the travel time between stations varies. In Fig. 10.1, the travel from S1 to S3 takes one time period, and

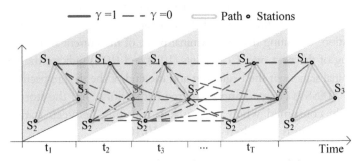

Figure 10.1 Temporal-spatial movement of MESS. *MESS*, Mobile energy storage system.

the travel from S1 to S2 takes two time periods. The red dashed and solid lines are the possible MESS mobility states. The red solid lines represent the actual mobility states. However, other mobility states are not realized because of the mobility constraints, and they are represented by the red dashed lines.

For example, there are three possible mobility states: S1−S1, S1−S2, and S1−S3 at the beginning of time period 1, and the MESS selects to stay at S1. Therefore S1−S1 is described by the solid line, while S1−S2 and S1−S3 are described by the dashed lines. The MESS mobility is described by the following constraints:

$$\sum_{(o,d),t_0 \in A_\tau} \gamma_{(o,d),t_0} = 1, \forall \tau \in T \tag{10.1}$$

$$\sum_{(o,d),t_0 \in A_{n,\tau_0,\text{in}}} \gamma_{(o,d),t_0} = \sum_{(o,d),t_0 \in A_{n,\tau_0,\text{out}}} \gamma_{(o,d),t_0}, \forall \tau \in T \tag{10.2}$$

$$\sum_{(o,d),t_0 \in A_{\text{dep},1,\text{out}}} \gamma_{(o,d),t_0} = 1 \tag{10.3}$$

$$\sum_{(o,d),t_0 \in A_{\text{ter},T,\text{in}}} \gamma_{(o,d),t_0} = 1 \tag{10.4}$$

where t_0 and τ_0 represent the beginnings of time intervals t and τ, respectively; (o,d) is the origin-destination station pair; $(o,d),t_0$ represents MESS mobility from the Station o to the Station d at t_0, A is the set of all possible MESS mobility strategies. $\gamma_{(o,d),t_0}$ is a binary variable for MESS mobility. When MESS mobility $(o,d),t_0$ is chosen, then $\gamma_{(o,d),t_0} = 1$; otherwise the mobility states are not realized, $\gamma_{(o,d),t_0} = 0$. Constraint (10.1) indicates that an MESS has a single mobility state at any time interval. Constraint (10.2) describes the traffic flow conservation, indicating that the terminal point of the current movement is the beginning of the next one. Constraints (10.3) and (10.4) indicate the initial and the terminal stations at the beginning and the end of the day, respectively.

The MESS is charged and discharged at stations, and the stored energy is partially consumed when the truck is traveling between stations. The MESS operation is described in (10.5)−(10.10).

$$\left(p_{e,i,t}^{\text{ch}} - p_{e,i,t}^{\text{di}}\right)^2 + \left(q_{e,i,t}^{\text{ess}}\right)^2 \leq \left(\gamma_{e,(i,i),t_0} S_e^{\text{ess}}\right)^2, \forall t \in T \tag{10.5}$$

$$p_{e,i,t}^{\text{ch}}, p_{e,i,t}^{\text{di}} \geq 0, \forall t \in T \tag{10.6}$$

$$p_{e,t}^{\text{travel}} = \left(1 - \sum_{i \in \textbf{Node}} \gamma_{e,(i,i),t_0}\right) p_e^{\text{travel}}, \forall t \in T \tag{10.7}$$

$$C_{e,t} = C_{e,0} + \sum_{\tau=1}^{t} \left(\eta \sum_{i \in \textbf{Node}} p_{e,i,\tau}^{\text{ch}} - \frac{1}{\eta} \sum_{i \in \textbf{Node}} p_{e,i,\tau}^{\text{di}} - p_{e,\tau}^{\text{travel}}\right) \Delta t, \forall t \in T \tag{10.8}$$

$$SOC_{e,\min} C_e^{\text{ess}} \leq C_{e,t} \leq SOC_{e,\max} C_e^{\text{ess}}, \forall t \in T \tag{10.9}$$

$$C_{e,T} = C_{e,0} \tag{10.10}$$

Constraints (10.5) and (10.6) describe the active and reactive power limits of MESS. Constraint (10.7) indicates that MESS consumes electricity when traveling. When MESS travels, $p_{e,t}^{\text{travel}} = p_e^{\text{travel}}$; otherwise, the MESS stays at charging stations, and the power consumption of MESS mobility is zero. Constraint (10.8) describes the MESS energy at time interval t. The electricity consumption for mobility in time period τ is $p_{e,\tau}^{\text{travel}} \cdot \Delta t$. If the MESS travels during time periods 1 and t, the battery's total electricity consumption is $\sum_{\tau=1}^{t} p_{e,\tau}^{\text{travel}} \cdot \Delta t$. Constraints (10.9) and (10.10) describe the SOC limits. The simultaneous charging and discharging constraints are excluded as they may lead to suboptimal solutions.

We linearize (10.5) via the following piecewise linearization technique, and the linearization accuracy is analyzed in [16]:

$$-\gamma_{e,(i,i),t_0} S_e^{\text{ess}} \leq p_{e,i,t}^{\text{ch}} - p_{e,i,t}^{\text{di}} \leq \gamma_{e,(i,i),t_0} S_e^{\text{ess}} \tag{10.11}$$

$$-\gamma_{e,(i,i),t_0} S_e^{\text{ess}} \leq q_{e,i,t}^{\text{ess}} \leq \gamma_{e,(i,i),t_0} S_e^{\text{ess}} \tag{10.12}$$

$$-\sqrt{2}\gamma_{e,(i,i),t_0} S_e^{\text{ess}} \leq p_{e,i,t}^{\text{ch}} - p_{e,i,t}^{\text{di}} + q_{e,i,t}^{\text{ess}} \leq \sqrt{2}\gamma_{e,(i,i),t_0} S_e^{\text{ess}} \tag{10.13}$$

$$-\sqrt{2}\gamma_{e,(i,i),t_0} S_e^{\text{ess}} \leq p_{e,i,t}^{\text{ch}} - p_{e,i,t}^{\text{di}} - q_{e,i,t}^{\text{ess}} \leq \sqrt{2}\gamma_{e,(i,i),t_0} S_e^{\text{ess}} \tag{10.14}$$

10.2.1.2 Distribution network constraints

The distribution power flow is described by the linear DistFlow Model [17] as follows:

$$\sum_{k \in \beta(i)} P_{ik,t} = P_{ji,t} + p_{i,t} \tag{10.15}$$

$$\sum_{k \in \beta(i)} Q_{ik,t} = Q_{ji,t} + q_{i,t} \tag{10.16}$$

$$U_{i,t} = U_{j,t} - \left(r_{ji}P_{ji,t} + x_{ji}Q_{ji,t}\right)/U_1 \tag{10.17}$$

$$p_{i,t} = G_{p,i,t} + p_{i,t}^{pv} - p_{i,t}^{ch} + p_{i,t}^{di} - p_{i,t}^{load} \tag{10.18}$$

$$q_{i,t} = G_{q,i,t} + q_{i,t}^{pv} + q_{i,t}^{ess} - q_{i,t}^{load} \tag{10.19}$$

$$0 \leq p_{i,t}^{pv} \leq p_{i,t}^{pv,output}, \forall t \in \mathbf{T} \tag{10.20}$$

$$\left(p_{i,t}^{pv}\right)^2 + \left(q_{i,t}^{pv}\right)^2 \leq \left(s_i^{pv}\right)^2, \forall t \in \mathbf{T} \tag{10.21}$$

$$S_{ji,min}^2 \leq P_{ji,t}^2 + Q_{ji,t}^2 \leq S_{ji,max}^2 \tag{10.22}$$

$$U_{i,min} \leq U_{i,t} \leq U_{i,max} \tag{10.23}$$

where U_1 is the slack node voltage magnitude. The decision variables include active and reactive line flows, nodal voltage magnitude, active and reactive power from the utility grid, and active and reactive power of PV and MESS units. Constraints (10.15)−(10.16) describe the power balance of distribution systems. Constraint (10.17) describes the voltage drop in lines. Constraints (10.18)−(10.19) are the nodal power injections. Constraints (10.20)−(10.21) describe the active and reactive power of PV units. Constraints (10.22)−(10.23) describe the operating limits of the system. Since the distribution network is radial, line flows and nodal voltage magnitudes are described as functions of nodal power injections, as stated below:

$$\mathbf{P}_t = \mathbf{A}^{-1}\left(\mathbf{G}_{p,t} + \mathbf{p}_t^{pv} - \mathbf{p}_t^{load} - \sum_e \mathbf{p}_{e,t}^{ch} + \sum_e \mathbf{p}_{e,t}^{di}\right), \forall t \in \mathbf{T} \tag{10.24}$$

$$Q_t = A^{-1}\left(G_{q,t} + q_t^{pv} - q_t^{load} + \sum_e q_{e,t}^{ess}\right), \forall t \in T \qquad (10.25)$$

$$U_t = U_1 + A^{-T}\left(rA^{-1}\left(p_t^{pv} - p_t^{load} - \sum_e p_{e,t}^{ch} + \sum_e p_{e,t}^{di}\right) + xA^{-1}\left(q_t^{pv} - q_t^{load} + \sum_e q_{e,t}^{ess}\right)\right), \forall t \in T \qquad (10.26)$$

where P_t, Q_t, and U_t are the vectors of $P_{i,t}$, $Q_{i,t}$, and $U_{i,t}$; A is the node-branch incidence matrix of the distribution network. A^{-1} represents the inverse of A, and A^{-T} represents the transposition and inverse of A; r and x are diagonal matrices of branch resistance and reactance. Finally, distribution systems with PV and MESS units are modeled by (10.1)−(10.4), (10.6)−(10.14), (10.20)−(10.23), and (10.24)−(10.26), where the MESS scheduling and active and reactive power of PV units are decision variables.

10.2.1.3 Compact form of the optimization model

The objective function of the proposed model is to minimize the electricity purchase cost from upstream grids, which is stated as follows:

$$C = \sum_{t \in T} c_p G_{p,t} \Delta t \qquad (10.27)$$

Since load demands in distribution systems are satisfied by the power from upstream grids and PV units, minimizing the electricity from upstream grids will facilitate PV power consumption.

Then, the distribution system operation with MESS and PV units is modeled in a compact formas follows:

$$\min_{x,y} \sum_{t \in T} c^T y_t \qquad (10.28)$$

$$\text{s.t.} \quad Mx_t + Ny_t + O\hat{\xi}_t \le a, \forall t \in T \qquad (10.29)$$

$$\sum_{\tau=1}^{t} By_\tau \le b, \forall t \in T \qquad (10.30)$$

$$\sum_{t=1}^{T} D_t x_t = d \qquad (10.31)$$

where $x_t \in \{0,1\}^{Nxt}$ is a binary variable vector representing the MESS mobility strategy; $y_t \in R^{Nyt}$ represents the power outputs of MESS and PV units; $\hat{\xi}_t$ represents PV power and load demand forecast. Constraint (10.29) represents (10.6), (10.7), (10.11)−(10.14), (10.20−10.23), and (10.24)−(10.26); constraint (10.30) represents (10.8−10.10); constraint (10.31) represents (10.1−10.4). Matrices M, N, O, B, and D and vectors a, b, c, and d are the corresponding coefficients.

10.2.2 Multistage robust optimization model

The PV power output, load demands, and MESS travel time are uncertain variables, and their uncertainty models are presented in this section. Subsequently, the multistage robust optimization model is formulated for the MESS scheduling considering the dynamically updated travel time and the nonanticipativity MESS operation constraints.

10.2.2.1 Model of uncertainty
10.2.2.1.1 Travel time
We determine the MESS scheduling strategy based on the actual travel time in the current time interval and the conservatively estimated travel time in the remaining time intervals to guarantee the feasibility of MESS mobility. The MESS is rescheduled at every time interval to diminish the influence of changing traffic conditions, as shown in Fig. 10.2.

The actual travel time is downloaded from the online traffic monitoring system, which is updated based on real-time traffic conditions. The conservative travel time is estimated based on historical data, as shown in Fig. 10.3. First, the historical travel time data is used to establish a cumulative distribution function (CDF). Then, the 95% quantile is selected as the

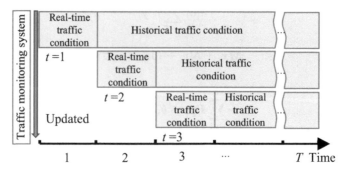

Figure 10.2 MESS scheduling based on real-time and historical traffic conditions. *MESS*, Mobile energy storage system.

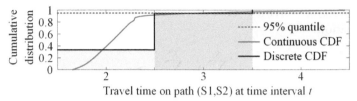

Figure 10.3 Cumulative probability distribution of travel time.

estimated travel time to guarantee the travel time reliability between stations. Note that we do not use the largest historical travel time because that would lead to overconservative decisions. Moreover, the travel time is discretized in 15-minute intervals, similar to that in the distribution system operation.

10.2.2.1.2 Photovoltaic power and load demand

The PV power $\xi_{p,t}$ and load demand $\xi_{d,t}$ are described by the following uncertainty sets with the upper and lower boundaries obtained from the historical data.

$$\Xi = \left\{ \begin{array}{l} \Xi_{p,t} = \left[\underline{\xi}_{p,t} \leq \xi_{p,t} \leq \overline{\xi}_{p,t} \right], \\ \Xi_{d,t} = \left[\underline{\xi}_{d,t} \leq \xi_{d,t} \leq \overline{\xi}_{d,t} \right], \\ \forall p \in C_{\mathrm{pv}}, \forall d \in C_{\mathrm{load}}, \forall t \in T \end{array} \right\} \tag{10.32}$$

The uncertainty set (10.32) may be considered conservative because it does not consider the correlation among random variables. In addition, certain random variables with a minor influence on system operation can be ignored. Here, we use principal component analysis (PCA) to reduce the number of random variables by mapping the original random variables to independent ones using the orthogonal transformation [18]. Then, we retain significant independent random variables with large variances. Hence, the uncertainty set of retained random variables is established as follows:

$$\Xi_t = \left\{ \Xi_{k,t} = \left[\underline{\xi}_{k,t} \leq \xi_{k,t} \leq \overline{\xi}_{k,t} \right], k = 1, \ldots, K \right\}, \quad \forall t \in T \tag{10.33}$$

where K and K' represent the numbers of retained and original random variables. Let $\lambda_1, \lambda_2, \ldots,$ and $\lambda_{K'}$ be the eigenvalues of the covariance matrix of the original variables. The number of retained random variables is determined based on the variance ratio ε, which is stated as

(10.34). The variance ratio ε is set at 95% to retain sufficient information.

$$\sum_{i=1}^{K} \lambda_i / \sum_{i=1}^{K'} \lambda_i \geq \varepsilon \qquad (10.34)$$

10.2.2.2 Multistage robust optimization problem

This section proposes a multistage robust optimization model for dynamic MESS scheduling considering uncertain PV power and load demands as well as the nonanticipation of MESS scheduling decisions. In Fig. 10.4, the current time interval is stage 1, and the remaining time intervals represent stages $2, \ldots, T$. The MESS scheduling is sequentially determined based on the realization of random variables up to the corresponding stage. Specifically, when determining the decisions at stage 1, the decisions at stage t ($t = 2, \ldots, T$) are not made, and their feasible regions are affected by those at stage 1. This is because the decisions at stage t depend on the realizations of random variables between stages 1 to t.

The uncertain travel time is modeled as follows: To obtain deterministic MESS mobility strategies at stage 1, the accurate travel time is assumed to be known at stage 1, which is estimated for the remaining stages. Accordingly, the deterministic MESS scheduling model (10.28)– (10.31) is expanded into a multistage robust optimization model (10.35)– (10.39). The MESS mobility strategy at stage 1 is determined before the realization of random variables. The MESS mobility strategies at the remaining stages and distribution system operations at all stages are made sequentially, considering the worst scenarios of uncertainty sets.

$$\min_{x_1} \max_{\xi_1 \in \Xi_1} \min_{y_1} \left\{ c^{\mathrm{T}} y_1 + \max_{\xi_2 \in \Xi_2} \min_{x_2, y_2} \left\{ c^{\mathrm{T}} y_2 + \cdots + \max_{\xi_T \in \Xi_T} \min_{x_T, y_T} c^{\mathrm{T}} y_T \right\} \right\}$$

$$(10.35)$$

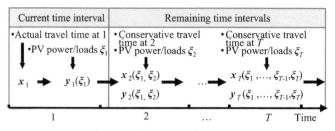

Figure 10.4 Decision procedure in the proposed multistage optimization.

$$\text{s.t.} \quad \boldsymbol{M}\boldsymbol{x}_1 + \boldsymbol{N}\boldsymbol{y}_1(\boldsymbol{\xi}_1) + \boldsymbol{O}\boldsymbol{\xi}_1 \leq a, \forall \boldsymbol{\xi}_1 \in \boldsymbol{\Xi}_1 \tag{10.36}$$

$$\boldsymbol{M}\boldsymbol{x}_t(\boldsymbol{\xi}_{[t]}) + \boldsymbol{N}\boldsymbol{y}_t(\boldsymbol{\xi}_{[t]}) + \boldsymbol{O}\boldsymbol{\xi}_t \leq a, \forall \boldsymbol{\xi}_t \in \boldsymbol{\Xi}_t, \forall t \in \boldsymbol{T}/\{1\} \tag{10.37}$$

$$\sum_{\tau=1}^{t} \boldsymbol{B}\boldsymbol{y}_\tau(\boldsymbol{\xi}_{[\tau]}) \leq b, \forall \boldsymbol{\xi}_\tau \in \boldsymbol{\Xi}_\tau, \forall t \in \boldsymbol{T} \tag{10.38}$$

$$\boldsymbol{D}_1\boldsymbol{x}_1 + \sum_{t=2}^{T} \boldsymbol{D}_t\boldsymbol{x}_t(\boldsymbol{\xi}_{[t]}) = d, \forall \boldsymbol{\xi}_t \in \boldsymbol{\Xi}_t \tag{10.39}$$

where $\boldsymbol{\xi}_t$ is the random variable vector at stage t; $\boldsymbol{\xi}_{[t]} =: \{\boldsymbol{\xi}_1, \boldsymbol{\xi}_2, \ldots, \boldsymbol{\xi}_t\}$ represents random variables from stages 1 to t. For $t = 2, \ldots, T$, $\boldsymbol{x}_t$ and $\boldsymbol{y}_t$ are determined based on random variables from stages 1 to t; thus, $\boldsymbol{x}_t$ and $\boldsymbol{y}_t$ are functions of random variables realized up to stage t. Model (10.35)−(10.39) is written in a compact form as follows:

$$\min_{\boldsymbol{x}_1} \quad \max_{\boldsymbol{\xi} \in \boldsymbol{\Xi}} \quad \min_{\{\boldsymbol{x}_{2,\ldots,T}, \boldsymbol{y}_{1,\ldots,T}\}} \sum_{t \in \boldsymbol{T}} c^{\mathrm{T}} \boldsymbol{y}_t(\boldsymbol{\xi}_{[t]}) \tag{10.40}$$

$$\text{s.t.} \quad \boldsymbol{M}\boldsymbol{x}_1 + \boldsymbol{N}\boldsymbol{y}_1(\boldsymbol{\xi}_1) + \boldsymbol{O}\boldsymbol{\xi}_1 \leq a, \forall \boldsymbol{\xi}_1 \in \boldsymbol{\Xi}_1 \tag{10.41}$$

$$\boldsymbol{M}\boldsymbol{x}_t(\boldsymbol{\xi}_{[t]}) + \boldsymbol{N}\boldsymbol{y}_t(\boldsymbol{\xi}_{[t]}) + \boldsymbol{O}\boldsymbol{\xi}_t \leq a, \forall \boldsymbol{\xi}_t \in \boldsymbol{\Xi}_t, \forall t \in \boldsymbol{T}/\{1\} \tag{10.42}$$

$$\sum_{\tau=1}^{t} \boldsymbol{B}\boldsymbol{y}_\tau(\boldsymbol{\xi}_{[\tau]}) \leq b, \forall \boldsymbol{\xi}_\tau \in \boldsymbol{\Xi}_\tau, \forall t \in \boldsymbol{T} \tag{10.43}$$

$$\boldsymbol{D}_1\boldsymbol{x}_1 + \sum_{t=2}^{T} \boldsymbol{D}_t\boldsymbol{x}_t(\boldsymbol{\xi}_{[t]}) = d, \forall \boldsymbol{\xi}_t \in \boldsymbol{\Xi}_t \tag{10.44}$$

After strategizing the MESS scheduling problem in the current time interval, the MESS scheduling problem in the next time interval is solved based on updated traffic conditions and uncertainty sets. Finally, MESS scheduling strategies for the entire period are obtained based on the dynamic optimization framework.

10.2.3 Solution methodology

The optimization problem (10.40)−(10.44) is intractable because of the infinite decision variables $\boldsymbol{x}_t(\boldsymbol{\xi}_{[t]})$ and $\boldsymbol{y}_t(\boldsymbol{\xi}_{[t]})$. This section provides a tractable formulation of (10.40)−(10.44) based on the affine decision rule and the duality theory.

10.2.3.1 Affine decision rule

In the proposed multistage optimization problem, decision variables $x_t(\xi_{[t]})$ and $y_t(\xi_{[t]})$ depend on random variables realized up to stage t. Because both continuous and binary decision variables exist in the optimization problem, the affine decision rule is used to approximate decision variables as affine functions of random variables to make the multistage decision problem tractable.

10.2.3.1.1 Linear decision rule

The linear decision rule describes continuous decision variables y_t as linear functions of random variables up to stage t. However, the direct application of the affine decision rule introduces large numbers of affine coefficients. Here, a simplified decision rule is used, in which the decision variables at stage t only depend on the random variables at stage t [19].

Accordingly, y_t are described as follows:

$$y_t(\xi_{[t]}) = Y_t \xi_t = y_0 + Y_{1,t}\xi_{1,t} + \cdots + Y_{K,t}\xi_{K,t}, Y_t \in R^{N_{yt} \times (K+1)} \quad (10.45)$$

Where $\xi_{0,t}$ is set as 1 without the loss of generality, Y_t is the matrix of affine coefficients for linear decision variables.

10.2.3.1.2 Binary decision rule

The affine function of binary decision variables x_t with respect to random variables is designed as follows [20]:

First, the range of each random variable $\xi_{k,t}$ is equally divided into $\rho + 1$ subintervals with the breakpoints $\mu_{r,k,t}$:

$$\mu_{r,k,t} = \underline{\xi}_{k,t} + r\left(\overline{\xi}_{k,t} - \underline{\xi}_{k,t}\right)/(\rho + 1), \quad r = 1, \ldots, \rho \quad (10.46)$$

Second, the piecewise binary function $G(\cdot)$ is defined as

$$G(\xi_t) = \left[1, G(\xi_{1,t}), \ldots, G(\xi_{K,t})\right]^T \quad (10.47)$$

where

$$G(\xi_{k,t}) = \left[G_1(\xi_{k,t}), \ldots, G_\rho(\xi_{k,t})\right], k = 1, \ldots, K \quad (10.48)$$

$$G_r(\xi_{k,t}) = 1, if \; \xi_{k,t} \geq \mu_{r,k,t}, r = 1, \ldots, \rho \quad (10.49)$$

Third, x_t are approximated as functions of the piecewise binary function $G(\xi_t)$:

$$x_t\left(\xi_{[t]}\right) = X_t G\left(\xi_t\right) \tag{10.50}$$

$$0 \leq X_t G\left(\xi_t\right) \leq 1, X_t \in \{0,\ 1\}^{N_{xt} \times (K\rho+1)} \tag{10.51}$$

where X_t is the matrix of affine coefficients for binary decision variables.

10.2.3.1.3 Lifted uncertainty set

Based on the linear and binary decision rules, the decision variables in the remaining stages are expressed as affine functions of random variables. Model (10.40)−(10.44) is converted into the following problem:

$$\min_{\{x_1, X_t, Y_t\}} \max_{\xi_t \in \Xi_t} \sum_{t \in T} c^{\mathrm{T}} Y_t \xi_t \tag{10.52}$$

$$\text{s.t. } Mx_1 + NY_1\xi_1 + O\xi_1 \leq a, \forall \xi_1 \in \Xi_1 \tag{10.53}$$

$$MX_t G\left(\xi_t\right) + NY_t\xi_t + O\xi_t \leq a, \forall \xi_t \in \Xi_t, \forall t \in T/\{1\} \tag{10.54}$$

$$\sum_{\tau=1}^{t} BY_\tau \xi_\tau \leq b, \forall \xi_\tau \in \Xi_\tau, \forall t \in T \tag{10.55}$$

$$D_1 x_1 + \sum_{t=2}^{T} D_t X_t G\left(\xi_t\right) = d, \forall \xi_t \in \Xi_t \tag{10.56}$$

However, it is still difficult to solve the optimization problem due to the binary function $G(\cdot)$. Hence, we define new random variables ξ'_t as follows:

$$\xi'_t = \left[\left[\xi'_{0,t}\right]^{\mathrm{T}}, \left[\xi'_{1,t}\right]^{\mathrm{T}}, \ldots, \left[\xi'_{K,t}\right]^{\mathrm{T}}\right]^{\mathrm{T}} = \left[[1,1], \left[\xi_{1,t}, G(\xi_{1,t})\right], \ldots,\right.$$

$$\left.\left[\xi_{K,t}, G\left(\xi_{K,t}\right)\right]\right]^T, \forall t \in T \tag{10.57}$$

Matrices L_1 and L_2 are introduced to transform the lifted random variable ξ_t' into ξ_t:

$$\begin{cases} \xi_t = L_1 \xi'_t \\ G(\xi_t) = L_2 \xi'_t \end{cases} \tag{10.58}$$

The lifted uncertainty set $\Xi'_{k,t}$ of $\xi'_{k,t}$ is discontinuous and nonconvex. Let the extreme points of $\Xi'_{k,t}$ be $(v_w, G(v_w))$, $w = 1,\ldots,2\rho + 2$, where v_w includes the upper and lower boundaries and breakpoints of a random variable $\xi_{k,t}$. Thus, the convex hull of $\Xi'_{k,t}$ is the convex combination of the extreme points [20], which is described as follows:

$$\text{conv}\left(\Xi'_{k,t}\right) = \left\{ \xi'_{k,t} = \left(\xi_{k,t}, G\left(\xi_{k,t}\right)\right)^\mathrm{T}, \exists \zeta_{w,t} \in R^+ : \right.$$

$$\left. \sum_{w=1}^{2\rho+2} \zeta_{w,t} = 1, \sum_{w=1}^{2\rho+2} \zeta_{w,t} v_w = \xi_{k,t}, \sum_{w=1}^{2\rho+2} \zeta_{w,t} G(v_w) = G\left(\xi_{k,t}\right) \right\} \tag{10.59}$$

The union of $\Xi'_{k,t}$ $(k = 1,\ldots,K)$ is presented as Ξ'_t, and the convex hull of Ξ'_t is denoted as $\text{conv}(\Xi'_t)$. Since $\text{conv}(\Xi'_t)$ is a close and bounded polyhedral set, it is represented in a compact form:

$$\text{conv}(\Xi'_t) = \left\{ \begin{array}{l} \xi'_t : \exists \zeta_t \in R^{K \times (2\rho+2)} : \\ A_{1,t}\xi'_t + A_{2,t}\zeta_t = h_t, \zeta_t \geq 0 \end{array} \right\} \tag{10.60}$$

where $A_{1,t}$, $A_{2,t}$, and h_t are coefficients obtained from (10.59).

10.2.3.2 Reformulation of the optimization Model
Based on the convex hull of the lifted uncertainty set that is presented in (10.60), the optimization problem (10.52)−(10.56) is reformulated as follows:

$$\min_{\{x_1, X_t, Y_t\}} \max_{\xi'_t \in \text{conv}(\Xi'_t)} \sum_{t \in T} c^\mathrm{T} Y_t L_1 \xi'_t \tag{10.61}$$

$$\text{s.t. } Mx_1 + NY_1 L_1 \xi'_1 + OL_1 \xi'_1 \leq a, \forall \xi'_1 \in \text{conv}(\Xi'_1) \tag{10.62}$$

$$MX_t L_2 \xi'_t + NY_t L_1 \xi'_t + OL_1 \xi'_t \leq a, \forall \xi'_t \in \text{conv}(\Xi'_t), \forall t \in T/\{1\} \tag{10.63}$$

$$\sum_{\tau=1}^{t} BY_\tau L_1 \xi'_\tau \leq b, \forall \xi'_\tau \in \text{conv}(\Xi'_\tau), \forall t \in T \tag{10.64}$$

$$D_1 x_1 + \sum_{t=2}^{T} D_t X_t L_2 \xi'_t = d, \forall \xi'_t \in \text{conv}(\Xi'_t) \tag{10.65}$$

The inner maximum problem of the objective function (10.61) is transformed into the minimum problem (10.66)−(10.67) based on the duality theory. Robust constraints (10.62)−(10.64) are transformed into tractable constraints (10.63)−(10.73) using the standard robust optimization technique. Constraint (10.65) is equivalently transformed into (10.74)−(10.75). Finally, the multistage robust MSS scheduling problem is transformed into the MILP (mixed integer linear programming) problem (10.66)−(10.75). The flowchart of the solution method is illustrated in Fig. 10.5.

$$\min_{\{x_1, X_t, Y_t, \pi_t, \omega_t, \sigma_{t,\tau}\}} \sum_{t \in T} h_t^{\mathrm{T}} \pi_t \tag{10.66}$$

$$\text{s.t. } A_{1,t}^{\mathrm{T}} \pi_t = (Y_1 L_1)^{\mathrm{T}} c, A_{2,t}^{\mathrm{T}} \pi_t \geq 0, \forall t \in T \tag{10.67}$$

$$h_1^{\mathrm{T}} \omega_1 \leq (a - Mx_1)^{\mathrm{T}} \tag{10.68}$$

$$A_{1,1}^{\mathrm{T}} \omega_1 = (NY_1 L_1 + OL_1)^{\mathrm{T}}, A_{2,1}^{\mathrm{T}} \omega_1 \geq 0 \tag{10.69}$$

$$h_t^{\mathrm{T}} \omega_t \leq a^{\mathrm{T}}, \forall t \in T/\{1\} \tag{10.70}$$

$$A_{1,t}^{\mathrm{T}} \omega_t = (MX_t L_2 + NY_t L_1 + OL_1)^{\mathrm{T}}, A_{2,1}^{\mathrm{T}} \omega_t \geq 0, \forall t \in T/\{1\} \tag{10.71}$$

$$\sum_{\tau=1}^{t} h_\tau^{\mathrm{T}} \sigma_{t,\tau} \leq b, \forall t \in T \tag{10.72}$$

$$A_{1,t}^{\mathrm{T}} \sigma_{t,\tau} = (BY_t L_1)^{\mathrm{T}}, A_{2,t}^{\mathrm{T}} \sigma_{t,\tau} \geq 0, 1 \leq \tau \leq t, \forall t \in T \tag{10.73}$$

$$D_1 x_1 - d = 0 \tag{10.74}$$

$$D_t X_t L_2 = 0, \forall t \in T/\{1\} \tag{10.75}$$

10.2.4 Case studies

The proposed method was tested on the augmented 33-bus [21] and 123-bus [22] distribution systems. The programs are developed in Matlab on a

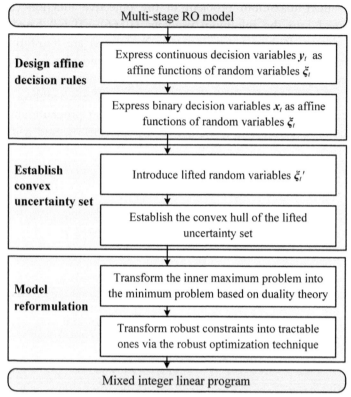

Figure 10.5 Flowchart of the solution method of the multistage robust optimization problem.

computer with a 2.20 GHz CPU and 64.0 GB RAM, and the optimization problems are solved using Yalmip combined with Gurobi.

Fig. 10.6 presents the 33-bus system along with the transportation network; there are one MESS and four stations in the system, which are marked with green. The locations of the charging stations are selected based on the global sensitivity analysis method by Lu et al. [23]. The maximum MESS charging or discharging power is 2 MW, with a capacity of 20 MWh. The MESS operational efficiency is 95%, and the MESS power consumption for mobility is 0.05 MW. The upper and lower limits of SOCs are designed as 80% and 20%, respectively, and the initial and final SOCs are set at 50%. The initial and terminal stations are S1, and the historical travel time data between charging stations are provided by the Gaode Maps Open Platform.

There are five PV units, each of which has a capacity of 2.5 MW. The load curve is that of the Southern California Edison and the solar

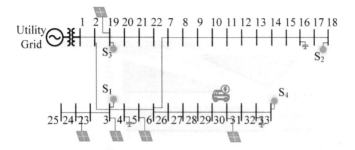

Figure 10.6 33-bus distribution network coupled with transportation network.

irradiance data of PV units are those of the National Renewable Energy Laboratory [24]. Fig. 10.7 shows the PV power curves on sunny and cloudy days. Compared with the PV power on sunny days, the average PV power is smaller on cloudy days, and the variance is larger since PV power is determined by solar irradiance.

10.2.4.1 Computation efficiency and optimization results with different uncertainty sets

The computation time and optimization results with PCA-based uncertainty sets are given in Table 10.1. On the one hand, compared with the original uncertainty sets, the computation time is much shorter for the PCA-based uncertainty sets. On the other hand, the optimization results vary for different uncertainty sets. Considering that the proposed optimization problem scale is relevant to the number of random variables, the PCA-based uncertainty set would significantly reduce the number of random variables and improve the computational efficiency of the optimization problem.

10.2.4.2 Mobile energy storage system operation for different photovoltaic power systems

Fig. 10.8 gives the MESS scheduling results on sunny and cloudy days. The MESS traveling and operating strategies are obtained by solving the proposed multistage optimization problem in the rolling-horizon optimization framework. In Fig. 10.8A, when there is no PV power in the morning, the MESS is discharged until reaching the lower SOC limit. The MESS is charged when PV units generate power. At 11:30, the PV power cannot be consumed due to the existing line power and nodal

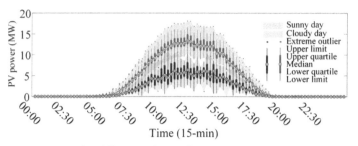

Figure 10.7 PV power for different solar irradiance conditions. *PV*, Photovoltaic.

Table 10.1 Optimization results and computation time.

Explained variance ratio	1	0.95
Number of random variables ξ_t	37	6
Computation time (s)	145.40	38.30
Daily operating cost ($)	4172.0	3926.4
Curtailed PV power (MWh)	46.91	48.19

voltage rise. So, MESS travels from S1 to S3 between 11:30 and 11:45 to store additional PV power. When the available PV power starts diminishing in the afternoon, MESS travels back to S1 between 16:00 and 16:15 and is discharged until 00:00. In Fig. 10.8B, the PV power declines on a cloudy day. The MESS is stationary in the initial charging station on a cloudy day because the benefit of MESS mobility is smaller than its mobility cost.

10.2.4.3 Mobile energy storage system operation in different traffic conditions

In the proposed MESS scheduling method, the travel time in the current time interval is real-time obtained, and the travel time in the following time intervals are conservative estimated. The actual travel time in the transportation network is given in Fig. 10.9A, which varies according to different paths. The vertical coordinates indicate paths, and the horizontal coordinates indicate departure times. The red and green colors indicate long and short travel times, respectively. Moreover, the travel time on the same path varies with traffic conditions. Fig. 10.9B shows the travel time on paths S3—S1 for different traffic conditions. In normal traffic, the actual travel time is relatively short. In congested traffic, the actual travel time is longer than the conservative travel time. The travel time between

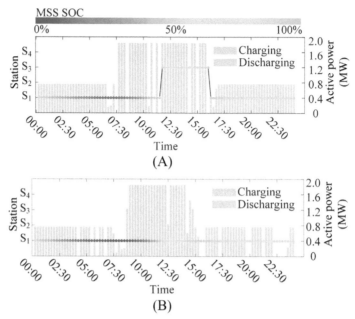

Figure 10.8 MESS schedule. (A) MESS schedule on a sunny day. (B) MESS schedule on a cloudy day. *MESS*, Mobile energy storage system.

charging stations is used to guide the MESS scheduling. Fig. 10.10 presents the MESS schedule in congested traffic, which remains the same before the traffic congestion. When path S3–S1 is congested, a new route is selected to avoid congested roads. Specifically, the MESS travels from S3 to S4 between 15:00 and 15:30 and from S4 to S1 between 16:00 and 17:00. Therefore the actual travel time reflects the traffic condition, and the rolling-horizon optimization with actual travel time provides reliable strategies for MESS mobility.

10.2.4.4 Comparison of scheduling methods
In the following cases, the proposed multistage MSS scheduling method is compared with other scheduling methods to verify its effectiveness.

Case 1: Proposed multistage MSS scheduling method.

Case 2: Day-ahead MSS scheduling using a two–stage robust optimization model that determines the MSS mobility strategy at the first stage and the operating strategy at the second stage.

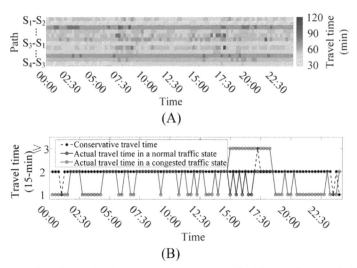

Figure 10.9 Travel time in the transportation network. (A) Travel time of different paths. (B) Travel time on paths S3−S1.

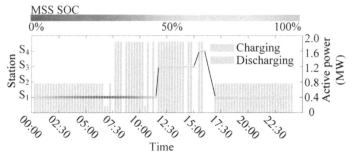

Figure 10.10 MESS schedule when road S3−S1 is congested. *MESS*, Mobile energy storage system.

Case 3: Distribution systems with stationary ESSs. The ESS locations are obtained using a two-stage robust optimization.

Case 4: Distribution systems without ESSs.

In Case 1, MSS mobility is strategized using the dynamic scheduling method with both actual and conservative travel times to guarantee the desired mobility strategy. The scheduling results for normal traffic and congested traffic are presented in Figs. 10.8 and 10.10. In Case 2, the day-ahead MSS scheduling strategy is obtained based on a two-stage robust optimization model. The day-ahead and intraday MSS scheduling results of Case 2 are given in Fig. 10.11. The intraday MSS scheduling is the actual operating result of the day-ahead strategy. The day-ahead MSS scheduling strategy might be infeasible in the intraday operation when the actual travel time varies, as shown in Fig. 10.11. The dashed black line represents the day-ahead MSS scheduling strategy, and the solid purple and green lines represent the actual intraday MSS movements in normal and congested traffic conditions, respectively. In Fig. 10.11A, the MSS traveling from S1 to S3 in the morning arrives 15 minutes earlier than expected. Then it arrives at S3 at 13:45, which is 15 minutes later than expected. In Fig. 10.11B, the MSS traveling from S3 to S1 in the afternoon arrives much later than expected. As shown in the figures, the day-ahead MSS mobility strategy is not fully implemented in the intraday operation since the actual travel time is different from the prediction. When the MSS arrives later than expected, the day-ahead scheduling strategy cannot be implemented in

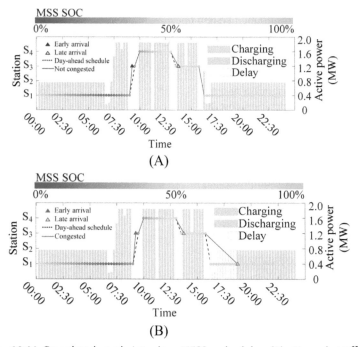

Figure 10.11 Day-ahead and intraday MESS schedule. (A) Normal traffic. (B) Congested traffic. *MESS*, Mobile energy storage system.

the intraday operation, thus affecting the economic benefit of the MSS operation. When the MSS arrives earlier than expected, it will be in idle mode, and its charging and discharging capacity will not be fully utilized.

To evaluate the performance of MSS and ESS scheduling methods, 1000 samples of PV power and loads are generated as actual scenarios, and the distribution system operating costs are calculated based on the generated scenarios. The computation time, operating costs, and PV power consumption are given in Table 10.2. In Case 1, the MSS is rescheduled based on the rolling-horizon optimization scheme, and the optimization time for each schedule is presented in the tables. The computation time of Case 4 is not given because there is no ESS in Case 4. Compared with the problems in Cases 2 and 3, the optimization model of Case 1 is a large-scale MILP problem, leading to a longer computation time. The computation time of Case 1 is acceptable considering that the time horizon of MSS scheduling is longer than the time of solving the optimization problem. The results of

Table 10.2 33-bus distribution system operation.

		Case 1	Case 2	Case 3	Case 4
Computation time (s)		48.19	6.10	4.05	—
Normal traffic	Distribution system operation cost ($)	3926.4	4202.3	4169.4	5700.1
	PV power consumption (MWh)	48.19	46.42	46.29	34.98
	MESS travel consumption (MWh)	0.025	0.1	—	—
Congested traffic	Distribution system operating cost ($)	4087.8	4365.1	4169.4	5700.1
	PV power consumption (MWh)	47.27	45.79	46.29	34.98
	MESS travel consumption (MWh)	0.0875	0.2375	—	—

MESS, Mobile energy storage system; PV, photovoltaic.

Cases 1—3 are better than those of Case 4, which indicates that ESSs can improve the economy of distribution system operation. The distribution system operating cost in Case 2 is larger than that in Case 3, although flexible and fixed positions are respectively considered in Cases 2 and 3. This is because the actual travel time may be different from that in the day-ahead estimate, leading to worse mobility strategies. In addition, the operating cost in Case 2 becomes even larger when traffic congestion occurs. The distribution system operating costs and PV power utilization of 1000 samples of Cases 1 and 2 are compared in Fig. 10.12. The operating cost of each sample is represented by an asterisk in the figure. If the asterisk is located at the diagonal line, the operating costs of Cases 1 and 2 are equal. As shown in Fig. 10.12A, the operating cost of each sample in Case 1 is smaller than that in Case 2. According to Fig. 10.12B, the PV power consumption of each sample in Case 1 is larger than that in Case 2. Therefore the proposed method outperforms the two-stage robust optimization model in all 1000 samples.

10.3 Data-driven routing and scheduling of mobile energy storage system for electricity arbitrage

10.3.1 Mobile energy storage system arbitrage problem

Fig. 10.13 shows the working principle of MESS arbitrage. The MESS travels among different charging stations that are connected to power

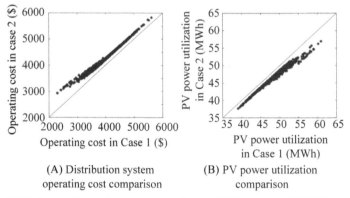

Figure 10.12 Comparison of operating results for 1000 samples. (A) Distribution system operating cost comparison. (B) PV power utilization comparison.

systems and located in transportation systems [25]. The MESS makes arbitrage revenue by charging at stations with low prices and discharging at those with high prices. Traffic time, electricity price, and battery operation influence the movement decisions and charging/discharging strategy of the MESS. Charging stations are connected to power grids, and charging/discharging prices are set as locational marginal prices (LMPs) of power grids. Assuming that the charging and discharging behaviors of the MESS do not affect the electricity prices, their impacts cannot be ignored when the power grid is penetrated by a large number of MESSs.

Therefore arbitrage revenue, travel costs, and battery capacity loss costs are considered in MESS arbitrage, with the objective function designed as

$$\max f = \sum_{t=1}^{96} \left(R_t - C_t^{\mathrm{tra}} \right) - C^{\mathrm{batt}} \tag{10.76}$$

The MESS operation constraints (10.1)−(10.10) are considered in the MESS arbitrage problem. Besides, the reactive power of the MESS is assumed to be zero, and the MESS is not assumed to consume electricity when traveling. The MESS is assumed to return to the starting station at the end of the day.

The arbitrage revenue of the MESS comes from the difference between selling and purchasing electricity.

$$R_t = \sum_{i \in Node} \gamma_{e,(i,i),t_0} p_{i,t} \left(\frac{1}{\eta_{\mathrm{di}}} p_{e,i,t}^{\mathrm{di}} - \eta_{\mathrm{ch}} p_{e,i,t}^{\mathrm{ch}} \right) \tag{10.77}$$

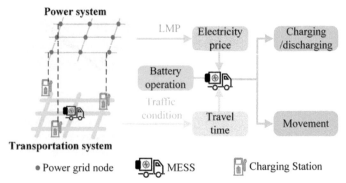

Figure 10.13 Working principle of MESS arbitrage. *MESS*, Mobile energy storage system.

where $p_{i,t}$ is the electricity price; η_{ch} and η_{di} are the charging and discharging efficiencies, and they are calculated using the model by Morstyn et al. [11]:

$$
\begin{cases}
\eta_{ch} = e_0 + e_1 SOC + e_2 SOC^2 + e_3 p_{e,i,t}^{ch} + e_4 p_{e,i,t}^{ch\,2} + e_5 p_{e,i,t}^{ch} SOC \\
\dfrac{1}{\eta_{di}} = h_0 + h_1 SOC + h_2 SOC^2 + h_3 p_{e,i,t}^{di} + h_4 p_{e,i,t}^{di\,2} + h_5 p_{e,i,t}^{di} SOC
\end{cases} \quad (10.78)
$$

The charging and discharging behaviors cause battery capacity degradation. The capacity loss E_{loss} is calculated using the nonlinear semiempirical degradation model by Xu et al. [12]:

$$
E_{loss} = \begin{cases}
E_{ini}\left[1 - \alpha_{sei}\exp\left(-\beta_{sei}\sum_j f_j^{d}\right) - (1 - \alpha_{sei})\exp\left(-\sum_j f_j^{d}\right)\right], E_{loss}' = 0 \\
E_{ini} - \left(E_{ini} - E_{loss}'\right)\exp\left(-\sum_j f_j^{d}\right), E_{loss}' > 0
\end{cases}
$$

$$(10.79)$$

where E_{loss}' and E_{loss} are the capacity losses before and after the period; α_{sei} and β_{sei} are the coefficients of the solid electrolyte interface film when the battery was formed; f_j^{d} is the capacity degradation of each charging cycle and is calculated using the model Xu et al. [12].

The degradation cost C^{batt} is calculated as

$$
C_{batt} = c_{batt} E_{loss} \tag{10.80}
$$

where c_{batt} is the cost coefficient.

The transportation cost caused by travel behaviors also reduces net profits:

$$
C_t^{tra} = \alpha_{tra} \sum_{(o,d),t_0 \in A_\tau} \gamma_{(o,d),t_0}, \quad \forall \tau \in T \tag{10.81}
$$

10.3.2 Deep reinforcement learning model

In the MESS arbitrage problem, the movement decision is discrete and the charging/discharging power is continuous. To cope with the discrete and continuous actions, a two-layer DRL model, consisting of the movement decision layer and the charging/discharging decision layer, is proposed to solve the MESS arbitrage problem. Two DRL agents are used to determine the actions of two layers, respectively. The movement decision agent selects the next charging station, and the charging/discharging decision agent determines the charging/discharging power at the current station.

10.3.2.1 State space
10.3.2.1.1 Movement decision layer
The state space contains the station where the MESS is staying, time periods at step i, SOC, traffic information, and electricity prices, as stated below:

$$s_i^m = \left\{ CS_i, t_i, SOC_i, \boldsymbol{T}_i^{tra}, \boldsymbol{p}_i^{arr} \right\} \qquad (10.82)$$

where CS_i is the station where the MESS stays at step i; t_i is the time period at step i; SOC_i is the SOC; $\boldsymbol{T}_i^{tra}$ is a vector representing the time of traveling from CS_i to all the stations; $\boldsymbol{p}_i^{arr}$ is a vector representing the predictive electricity prices of all the stations when the MESS arrives.

10.3.2.1.2 Charging/discharging decision layer
The agent makes decisions according to the actual electricity prices of current stations and time slots. Besides, the SOC is an important factor for decision-making. Hence, the state space is designed as

$$s_i^{ch} = \left\{ CS_i, t_i, SOC_i, p_i^{stay} \right\} \qquad (10.83)$$

10.3.2.2 Action space
10.3.2.2.1 Movement decision layer
The action is the next destination CS_i^{next}, as stated below:

$$a_i^m = \left\{ CS_i^{next} \right\} \qquad (10.84)$$

10.3.2.2.2 Charging/discharging decision layer
The action is the charging or discharging of power, as stated below:

$$a_i^{ch} = \left\{ P_i^{ch} \right\} \qquad (10.85)$$

10.3.2.3 Reward function
10.3.2.3.1 Movement decision layer
The reward function includes both revenues and costs from the movement and charging/discharging of the MESS. Specifically, the positive reward is the revenue of charging/discharging, and the negative reward includes the travel cost and the punishment of constraint violation, as stated below:

$$r_i^m = r_i^{ch} - C_i^{tra} - r_i^{m,\,P} a_i^{ch} = \left\{ P_i^{ch} \right\} \qquad (10.86)$$

where r_i^{ch} is the revenue of charging/discharging, which is the reward function of the charging/discharging decision layer; C_i^{tra} is the travel cost per period, which is calculated as

$$C_i^{tra} = \alpha_{tra} T_i^{tra} \left(CS_i^{next} \right) \tag{10.87}$$

where $T_i^{tra}(CS_i^{next})$ is the travel time from CS_i to CS_i^{next}.

$r_i^{m,p}$ is the punishment cost for failure to return to the starting station.

$$r_i^{m,p} = \begin{cases} \alpha_{tra} T_i^{tra}(CS_{start}), & t_i = 96 \text{ and } CS_i \neq CS_{start} \\ 0, \textit{otherwise} \end{cases} \tag{10.88}$$

where CS_{start} is the start station; $T_i^{tra}(CS_{start})$ is the travel time from CS_i to CS_{start}.

10.3.2.3.2 Charging/discharging decision layer

The reward is designed as arbitrage profits minus battery degradation costs and punishment costs of constraint violation, as stated below:

$$r_i^{ch} = p_i^{stay} P_i^{ch} - r_i^{ch,p} - C_i^{batt} \tag{10.89}$$

where $r_i^{ch,p}$ is the punishment caused by violating constraint (10.10):

$$r_i^{ch,p} = \begin{cases} \alpha_{SOC} |SOC_{i+1} - SOC_{init}|, & t_i = 96 \\ 0, \textit{otherwise} \end{cases} \tag{10.90}$$

To calculate the immediate battery degradation cost, the degradation cost per unit of power is estimated by:

$$\alpha_{batt} = c_{batt} \frac{E_{re,d}^{start} - E_{re,d}^{end}}{\sum_{i=1}^{T_d} |P_i^{ch}|} \tag{10.91}$$

where α_{batt} is the unit cost; $E_{re,d}^{start}$ is the remaining capacity at the beginning of the d_{th} calculation; $E_{re,d}^{end}$ is the remaining capacity at the end of the d_{th} calculation; T_d indicates that α_{batt} is updated every T_d period.

The degradation cost of each charging/discharging action is:

$$C_i^{batt} = \alpha_{batt} |P_i^{ch}| \tag{10.92}$$

10.3.2.4 Transition function

The executed actions determine transitions of environment states. The transitions of current charging stations, time periods, and SOC are stated as

$$CS_{i+1} = d_i^m \tag{10.93}$$

$$t_{i+1} = \begin{cases} t_i + 1, \text{ if } CS_i \neq a_i^m \\ t_i + T_i^{tra}\left(a_i^m\right), \text{ otherwise} \end{cases} \tag{10.94}$$

$$SOC_{i+1} = \begin{cases} SOC_i - \eta_i^{ch} \cdot \dfrac{a_i^{ch} \cdot \Delta t}{E_{ini}}, a_i^{ch} \leq 0 \\[4mm] SOC_i - \dfrac{1}{\eta_i^{dis}} \cdot \dfrac{a_i^{ch} \cdot \Delta t}{E_{ini}}, a_i^{ch} > 0 \end{cases} \tag{10.95}$$

10.3.3 Implementation

Rainbow and TD3 algorithms are used in the movement and charging/discharging decision layers, respectively, because of their discrete and continuous spaces. The implementation process of DRL is divided into training and application stages. In the training stage, the DRL agents learn policies gradually by interacting with environments and updating the parameters of networks. In the application stage, the DRL agents determine actions to acquire a high reward based on learned policies. Fig. 10.14 shows the flowchart of the training and application stages.

10.3.3.1 Training stage

There are two agents in the two-layer DRL model. To ensure sufficient training of both agents, a sequential-training method is proposed to train DRL models. Specifically, the charging/discharging decision agent is trained in advance using the TD3 algorithm. The mapping relationship from states to actions is fitted by interacting with environments, and the parameters of critic networks are optimized by minimizing the loss function. To guarantee the diversity of states, the station of each period is randomly generated when training the charging/discharging decision agent. The movement decision agent is trained using the Rainbow algorithm based on the pretrained charging/discharging decision agent. The movement decision agent selects transitions with the priority to approximate the distribution of return, and the parameters of dueling networks are updated by minimizing the loss function. After the training stage, the parameters of networks are frozen, and both agents have learned policies.

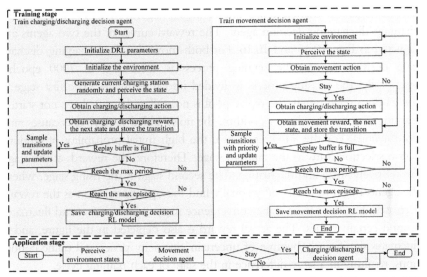

Figure 10.14 Training and application processes of the two-layer DRL model. *DRL,* Deep reinforcement learning.

10.3.3.2 Application stage

In the application stage, the two DRL agents take action sequentially. First, the DRL agent in the movement decision layer perceives environments and chooses the next destinations of MESS mobility. Then, the DRL agent in the charging/discharging layer takes action only if the DRL agent in the movement decision layer determines that the MESS is stationary. The agents make decisions of one period at every step and obtain the decisions of a day in a rolling-horizon way.

10.3.4 Case studies

The proposed method is tested using the electricity and traffic information of 31 charging stations in San Diego, USA. The charging/discharging prices of charging stations are designed as LMPs, which are available on the website of the California Independent System Operator [26]. The traffic information between different stations is downloaded from the HERE Maps API [27]. Table 10.3 gives the MESS parameters.

10.3.4.1 Training process analysis

Based on the designed training strategy, the charging/discharging decision agent is trained using the TD3 algorithm, and subsequently, the movement decision agent is trained using the Rainbow algorithm and the pretrained charging/discharging decision agent. The reward curves of the two agents are depicted in Figs. 10.15 and 10.16. For both movement and charging/discharging decision agents, the training process is converged in 7000 epochs. Specifically, the training process is divided into three stages. The first stage is the random stage where the replay pool is not full and agents have not started learning. For the Rainbow algorithm, the noise added to networks causes random actions. For the TD3 algorithm, the high initial probability of random action selection leads to the random stage. Therefore, the rewards at this stage are relatively small and fluctuating. The second stage is the rising stage, where the agents begin to learn and gradually find the optimal solution as the reward increases. The third stage is the convergence stage, where the reward fluctuates around certain values. The actual reward is also depicted in the figure, and it fluctuates significantly because of uncertain electricity prices and traffic conditions, resulting in a large difference in the state of each training cycle.

Table 10.3 Mobile energy storage system parameter.

Parameter	Value	Parameter	Value
Initial capacity (MWh)	15	Maximum charging power (MW)	3
Max SOC	1	Maximum discharging power (MW)	3
Min SOC	0.2	Period (min)	15
e_0	1.00	e_1	4.00×10^{-3}
e_2	-3.11×10^{-3}	e_3 (kW^{-1})	-1.59×10^{-5}
e_4 ((kW)$^{-2}$)	3.40×10^{-10}	e_5 (kW^{-1})	3.22×10^{-6}
h_0	1.00	h_1	-4.60×10^{-3}
h_2	4.13×10^{-3}	h_3 (kW^{-1})	1.67×10^{-5}
h_4 ((kW)$^{-2}$)	4.70×10^{-10}	h_5 (kW^{-1})	-4.53×10^{-6}

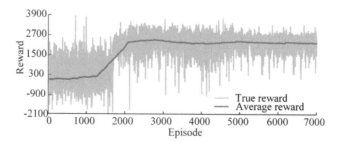

Figure 10.15 Reward curve of charging/discharging decision agents.

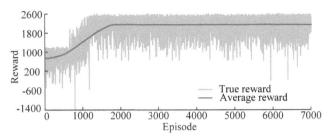

Figure 10.16 Reward curve of movement decision agents.

10.3.4.2 Application process analysis
10.3.4.2.1 Performance analysis for one day

Figs. 10.17 and 10.18 depict the MESS routing and scheduling in different stations, where CS 1, CS 2, ..., and CS 31 refer to 31 charging stations, and CS 0 indicates that the MESS is traveling. As shown in the figures, the MESS starts its trip at CS 15 and then arrives at CS 16, CS 31, and CS 22. Because of the electricity price changes, the MESS travels to stations with higher prices for power discharging and to those with lower prices for power charging. Finally, it returns to CS 15 at the end of the day. When the price changes are minor, the MESS neglects small arbitrage profits and is operated in idle mode, thus reducing travel costs and extending battery life. In some periods, the MESS is neither charged at the station with the lowest price nor discharged at with the highest price. This is because electricity prices and travel costs are comprehensively considered in the designed DRL model. When the price differences between different stations are small, arbitrage revenues may not cover travel costs, although there still exists space for arbitrage.

10.3.4.2.2 Performance analysis under fluctuating prices

The fluctuation of charging/discharging prices is important to MESS arbitrage. Generally, there is great arbitrage potential under highly fluctuating electricity prices. To demonstrate that the Rainbow-TD3 method makes proper decisions in different price conditions, a day with highly fluctuating prices is selected to test the performance of the proposed method. As shown in Fig. 10.19, the MESS locations are frequently changed, because the MESS not only affords the costs of frequent travels but also obtains high profits owing to the price fluctuation. Fig. 10.20 sketches the scheduling results that show that the MESS is charged or discharged more times with more power to improve arbitrage revenue.

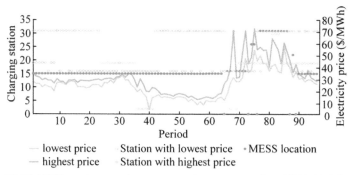

Figure 10.17 MESS routing result among stations on one day. *MESS*, Mobile energy storage system.

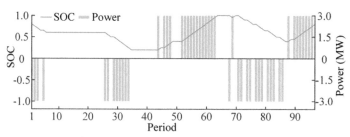

Figure 10.18 SOC and charging/discharging power of MESS on one day. *MESS*, Mobile energy storage system; *SOC*, state of charge.

10.3.4.2.3 Performance analysis under different traffic conditions

To verify that the proposed method generates well-organized routing strategies, we set all the roads to CS 31 congested by increasing travel costs by 1.5 times. Fig. 10.21 shows the routing and scheduling results in normal and congested traffic conditions. In the congested traffic condition, although all the roads to CS 31 are congested, MESS still selects CS 31 to discharge power because of its great arbitrage potential. To reduce travel costs to CS 31, Route $15 \rightarrow 30$ and Route $30 \rightarrow 31$ are selected to replace the original routing plans, Routes $15 \rightarrow 16$, $16 \rightarrow 24$, and $24 \rightarrow 31$. The charging and discharging strategies are also changed during periods 74 and 76 due to route changes.

10.3.4.3 Comparative analysis with model-based methods

To verify the effectiveness of the proposed data-driven method, the model-based MESS arbitrage method is tested for comparison, as shown

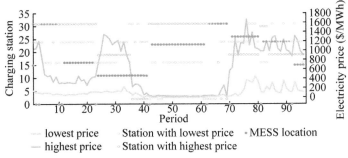

Figure 10.19 MESS routing strategy under fluctuating prices. *MESS*, Mobile energy storage system.

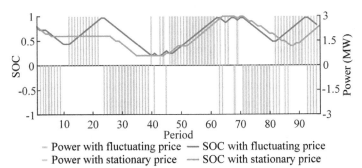

Figure 10.20 MESS charging/discharging strategies on days with different prices. *MESS*, Mobile energy storage system.

in Table 10.4. The performances of the two methods are analyzed in two cases: 84-scenario and historical-data cases. In the 84-scenario cases, representative scenarios are used to simulate the actual situations of MESS operation in the whole life cycle. For the model-based method, the MESS arbitrage is cast as MILP problems, which are solved by Yalmip combined with Gurobi. Method 1 is the theoretical benchmark with the ideal charging/discharging efficiency, and other model-based methods consider actual charging/discharging efficiencies. The proposed method is tested in the 84-scenario cases to verify its ability to handle uncertainties. Besides, its results in the historical-data case are also given.

Table 10.5 gives the profits obtained by different methods. In Case 1, the proposed method (Method 5) achieves the highest life cycle profit and relatively high annual profits. On the one hand, although the proposed method obtains smaller annual profits than the benchmark (Method 1) and Method 2, the differences are minor, which are 3.26% and 0.65%, respectively.

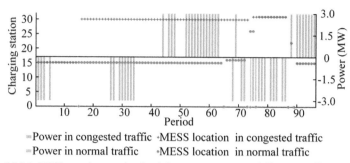

Figure 10.21 MESS routing and scheduling strategies in different traffic conditions. *MESS*, Mobile energy storage system.

Compared with Method 1, Method 2 exhibits a closer annual revenue to the proposed method, which indicates that similar battery models narrow the gap between model-based and DRL methods. On the other hand, Methods 1 and 2 have actual information on electricity prices and travel times of all the periods, which is impossible to acquire in real-world scenarios. Hence, the proposed method uses less information to obtain high profits. Furthermore, the proposed method achieves the longest battery life and the highest life cycle profits, because large weighting coefficients are designed in the reward function. Hence, DRL agents tend to delay immediate arbitrage profits in exchange of long-term benefits, which means that the Rainbow-TD3 method balances battery life and arbitrage profits. Besides, Method 3 shows the worst performance due to neglecting battery degradation.

Predictive and representative scenarios are two common ways of handling uncertainties. Due to uncertain electricity prices and traffic conditions, it is hard to make accurate predictions in actual situations, which explains the difference between Methods 1 and 4 in Case 1. Method 4 has full knowledge of predictive information, while the proposed method (Method 5) only knows the information of current periods and the predictive information of the next two periods. Nevertheless, Method 5 obtained better results than Method 4, which indicates that the Rainbow-TD3 method has a stronger ability to handle uncertainties than those methods using predictive information. Representative scenarios are adopted in Case 1, and the historical data is adopted in Case 2 to generate MESS routing and scheduling strategies. As shown in Table 10.5, Method 6 in Case 2 outperforms benchmarks in Case 1. Therefore the Rainbow-TD3 method has better performance than model-based methods with predictive or representative scenarios.

Table 10.4 Testing case description.

Case	No.	Method	Degradation	Charging/ discharging efficiency	Information awareness
Case 1: 84 scenarios	1	Model-based	Considered	1.0	Actual information of all the periods
	2	Model-based	Considered	Changing	Actual information of all the periods
	3	Model-based	—	Changing	Actual information of all the periods
	4	Model-based	Considered	Changing	Predictive information of all the periods
	5	Rainbow-TD3	Considered	Changing	Actual information of current periods and predictive prices of the next two periods
Case 2: Historical data	6	Rainbow-TD3	Considered	Changing	Actual information of current periods and predictive prices of the next two periods

Table 10.5 Results obtained by different methods.

Case	Method no.	Battery life (day)	Total profit ($)	Average annual profit ($/year)	Time (s/day)
Case 1: 84 scenarios	1	1830	1,856,561.11	370,297.71	1194.43
	2	1800	1,780,081.79	360,961.03	1782.53
	3	1110	560,702.96	184,375.30	1782.53
	4	1770	1,471,464.40	303,437.57	1782.53
	5	1890	1,856,973.31	358,621.83	0.10
Case 2: Historical data	6	1950	2,188,881.41	409,713.70	0.10

The computation time is given in the last column of Table 10.5. In the application stage, Rainbow–TD3 obtains daily strategies in about 0.1 second while model-based methods are more time-consuming. This is because the proposed method has learned good strategies and the deep networks have established the mapping from states to appropriate actions after a long-time training stage. In the application stage, with the learned policies, the proposed method obtains MESS routing and scheduling results through fast mapping enabled by deep networks. However, the model-based methods have to solve problems from scratch for the schedules of each day. Besides, a large number of integer variables also increase the computational burden of model-based methods. Therefore the proposed method achieves higher computational efficiency in the application stage.

10.4 Conclusion

This chapter proposes a dynamic MESS scheduling method for PV power integration that considers the coupling of transportation and power distribution networks and explores the arbitrage strategies considering complicated battery models.

To hedge against the variable traffic conditions, MESS is rescheduled at every time interval using both real time and conservative travel time to guarantee the desired mobility strategy. The multistage robust optimization problem is established considering the nonanticipativity MESS operation constraints as well as the uncertain PV power and load demands. The proposed multistage robust optimization model is transformed into a tractable MILP problem based on the affine decision rule and duality theory. The proposed method is tested on the augmented 33-bus distribution system. Simulation results indicate that PCA-based uncertainty sets capture the PV power and load demands correlations to reduce the number of random variables, which significantly improves the computational efficiency of the affine decision rule-based multistage robust optimization. Compared with the day-ahead MESS and stationary ESS schedules, the dynamic MESS operation strategy adapts to the variable traffic conditions, which provides flexible MESS routing and scheduling strategies and improves large-scale renewable power integration into distribution systems.

This chapter proposes a data-driven uncertainty-adaptive method for MESS arbitrage considering the complicated operational characteristics of batteries. To achieve discrete mobility and continuous power transmission, the two-layer DRL method is proposed to obtain MESS routing and scheduling decisions. A sequential training process is proposed to accelerate the convergence of DRL model training. The proposed method is tested using real-world electricity and traffic data from charging stations. Simulation results show that the proposed method obtains movement and charging/discharging decisions in line with MESS operation rules and battery capacity degradation characteristics. Compared with traditional model-based methods that require complete and accurate future information, the proposed method obtains high arbitrage profits by learning arbitrage strategies from historical data and making effective decisions with limited information. Besides, the proposed method captures the uncertain factors of electricity prices and traffic conditions from historical data, thus outperforming those with predictive or representative scenarios in uncertain environments.

References

[1] Nazemi M, Dehghanian P, Lu X, Chen C. Uncertainty-aware deployment of mobile energy storage systems for distribution grid resilience. IEEE Transactions on Smart Grid 2021;12(4):3200−14.
[2] Ghasemi S, Moshtagh J. Distribution system restoration after extreme events considering distributed generators and static energy storage systems with mobile energy storage systems dispatch in transportation systems. Applied Energy 2022;310:118507.
[3] Kim J, Dvorkin Y. Enhancing distribution system resilience with mobile energy storage and microgrids. IEEE Transactions on Smart Grid 2019;10(5):4996−5006.
[4] Erenoğlu AK, Erdinç O. Post-Event restoration strategy for coupled distribution-transportation system utilizing spatiotemporal flexibility of mobile emergency generator and mobile energy storage system. Electric Power Systems Research 2021;199: 107432.

[5] Khodayar ME, Wu L, Li Z. Electric vehicle mobility in transmission-constrained hourly power generation scheduling. IEEE Transactions on Smart Grid 2013;4(2):779−88.

[6] Ebadi R, Yazdankhah AS, Kazemzadeh R, Mohammadi-Ivatloo B. Techno-economic evaluation of transportable battery energy storage in robust day-ahead scheduling of integrated power and railway transportation networks. International Journal of Electrical Power & Energy Systems 2021;126(Part A):106606.

[7] Abdeltawab HH, Mohamed YAI. Mobile energy storage scheduling and operation in active distribution systems. IEEE Transactions on Industrial Electronics 2017;64(9):6828−40.

[8] Zhai Z, Zhou Y, Li X, Wu J, Xu Z, Xie X. Nonanticipativity and all-scenario-feasibility: state of the art, challenges, and future in dealing with the uncertain load and renewable energy. Proceedings of the CSEE 2020;40(20):11−26.

[9] Cobos NG, Arroyo JM, Alguacil N, Wang J. Robust energy and reserve scheduling considering bulk energy storage units and wind uncertainty. IEEE Transactions on Power Systems 2018;33(5):5206−16.

[10] Wang W, Wu Y. Is uncertainty always bad for the performance of transportation systems? Communications in Transportation Research 2021;1:100021, ISSN 2772−4247.

[11] Morstyn T, Hredzak B, Aguilera RP, Agelidis VG. Model predictive control for distributed microgrid battery energy storage systems. IEEE Transactions on Control Systems Technology 2018;26(3):1107−14.

[12] Xu B, Oudalov A, Ulbig A, Andersson G, Kirschen DS. Modeling of lithium-ion battery degradation for cell life assessment. IEEE Transactions on Smart Grid 2018;9(2):1131−40.

[13] Peng B, Keskin MF, Kulcsár B, Wymeersch H. Connected autonomous vehicles for improving mixed traffic efficiency in unsignalized intersections with deep reinforcement learning. Communications in Transportation Research 2021;1:100017, ISSN 2772−4247.

[14] ChenX., QuG., TangY., LowS., LiN. Reinforcement learning for selective key applications in power systems: recent advances and future challenges. IEEE Trans. Smart Grid. Available from: https://doi.org/10.1109/TSG.2022.3154718.

[15] Lu Z, Xu X, Yan Z, Shahidehpour M. Multistage robust optimization of routing and scheduling of mobile energy storage in coupled transportation and power distribution networks. IEEE Transactions on Transportation Electrification 2022;8(2):2583−94. Available from: https://doi.org/10.1109/TTE.2021.3132533.

[16] Chen X, Wu W, Zhang B. Robust restoration method for active distribution networks. IEEE Transactions on Power Systems 2016;31(5):4005−15.

[17] Song Y, Zheng Y, Liu T, et al. A new formulation of distribution network reconfiguration for reducing the voltage volatility induced by distributed generation. IEEE Transactions on Power Systems 2020;35(1):496−507.

[18] Ning C, You F. Data-driven decision making under uncertainty integrating robust optimization with principal component analysis and kernel smoothing methods. Computers & Chemical Engineering 2018;112:190−210.

[19] Zhou Y, Shahidehpour M, Wei Z, Sun G, Chen S. Multistage robust look-ahead unit commitment with probabilistic forecasting in multi-carrier energy systems. IEEE Transactions on Sustainable Energy 2021;12(1):70−82.

[20] Bertsimas D, Georghiou A. Binary decision rulesfor multistage adaptive mixed-integer optimization. Mathematical Programming 2018;167(2):395−433.

[21] Ding T, Li C, Yang Y, Jiang J, Bie Z, Blaabjerg F. A two-stage robust optimization for centralized-optimal dispatch of photovoltaic inverters in active distribution networks. IEEE Transactions on Sustainable Energy 2017;8(2):744−54.

[22] Wang X, Li Z, Shahidehpour M, Jiang C. Robust line hardening strategies for improving the resilience of distribution systems with variable renewable resources. IEEE Transactions on Sustainable Energy 2019;10(1):386−95.

[23] Lu Z, Xu X, Yan Z, Wang H. Density-based global sensitivity analysis of islanded microgrid loadability considering distributed energy resource integration. Journal of Modern Power Systems and Clean Energy 2019;8(1):94–101.

[24] Lu Z, Xu X, Yan Z. Data-driven stochastic programming for energy storage system planning in high PV-penetrated distribution network. International Journal of Electrical Power & Energy Systems 2020;123:106326.

[25] Chen T, Xu X, Wang H, Yan Z. Routing and scheduling of mobile energy storage system for electricity arbitrage based on two-layer deep reinforcement learning. In: IEEE Transactions on Transportation Electrification, 2022. Available from: https://doi.org/10.1109/TTE.2022.3201164.

[26] California Independent System Operator. Open Access Same time Information System (OASIS) price reports. Available from: http://oasis.caiso.com/mrioasis/logon.do.

[27] Here developer. Here routing API. Available from: https://developer.here.com/documentation/routing-api/api-reference-swagger.html.

CHAPTER 11

Application of Internet of Energy and digitalization in smart grid and sustainability

Yuekuan Zhou[1,2,3,4]

[1]Sustainable Energy and Environment Thrust, Function Hub, The Hong Kong University of Science and Technology (Guangzhou), Nansha, Guangdong, P.R. China
[2]Department of Mechanical and Aerospace Engineering, The Hong Kong University of Science and Technology, Clear Water Bay, Hong Kong SAR, P.R. China
[3]HKUST Shenzhen-Hong Kong Collaborative Innovation Research Institute, Futian, Shenzhen, P.R. China
[4]Division of Emerging Interdisciplinary Areas, The Hong Kong University of Science and Technology, Clear Water Bay, Hong Kong SAR, P.R. China

11.1 Introduction

With four industrial revolutions, communication has transitioned from human–to–human verbal communication to human–to–machine communication [1], together with the transformation from a things-based to a services-based energy framework. Renewable energy penetration in smart grids with advanced communication technologies requires interaction among prosumers [2]. To evaluate the future integrated multidimensional system, an indicator-based approach that includes expert elicitation can identify features as much as possible considering future changes and interdependencies between energy systems [3]. Industry 4.0 requires information, digitalization, and sustainability [4]. The development of smart cities requires the combined development of smart homes, intelligent transportation, advanced manufacturing, and self-powered sensors with a 5 G-powered Internet of Things (IoT) to accelerate digitalization in smart cities [5].

The fast development of smart sensors and metering can transform energy digitalization. Digital technologies can improve accuracy and efficiency in energy performance prediction, energy efficiency through synergistic integration, and operation. Inderwildi et al. [6] comprehensively reviewed big data, machine learning, and IoT in the decarbonizationof integrated energy systems. However, advanced policies are necessary to guarantee resilience and energy security.

Advances in Digitalization and Machine Learning for Integrated
Building-Transportation Energy Systems
DOI: https://doi.org/10.1016/B978-0-443-13177-6.00010-2

211

The twin transitions of decarbonization and digitalization show a pace mismatch, in which the development pace of digitalization is much faster than that of decarbonization [7]. Therefore synergies and complementarities between decarbonization and digitalization are quite necessary. Advancement in digitalization can promote decarbonization concerning technological advancement (such as smart sensing/metering, fast and efficient communication and controls [8], fast response [9], etc.), optimal energy structure, capital cost saving of labor, and end-user participation. Based on evidence from China, Wu et al. [10] concluded that the fast development of the Internet can enhance energy saving and emission reduction. Haldar and Sethi [11] quantitatively analyzed the impact of information and communication technology on carbon emissions. According to the environmental Kuznets curve, the development of the economy can be promoted using the Internet, which can reduce emissions and promote decarbonization [12]. Sareen [13] studied the social acceptance of digitalization in the energy transition to eliminate energy inequality.

In this chapter, an overview of Internet of Energy (IoE) and digitalization has been introduced. The correlation between digitalization and sustainability has been uncovered, together with their applications in energy networks and power grids. Based on the current development status, frontier guidelines are provided for policymakers, designers, and operators for energy internet (EI) networks.

11.2 Energy digitalization and Internet of Energy

In the EU digital targets for 2030 [14], step-by-step actions were taken, i.e., public consultation, feedback, and stakeholder engagement. Ren et al. [15] studied the correlations between energy consumption and Internet development. Evidence indicates that the fast development of the Internet will reduce energy consumption. By developing the regression model, Hao et al. [16] quantified correlations between digitalization and electricity consumption based on data samples from 30 Chinese provinces from 2006 to 2018. Results indicated that Internet development will negatively reduce power consumption, especially with the further development of finance, education level, industrial structure, and technologies. Zhang et al. [17] studied the role of energy storage digitalization in sustainability transformation. Results showed that digitalization can contribute to technological innovation with the IoT in fundamental roles.

The IoT can drive energy digitalization, and blockchain decentralizes the transactive EI [18]. Digitalization and decentralization in EI heavily rely on technologies and infrastructures. In the context of transformation toward the data digitization era, Tayal [19] applied data analytics and machine learning to quantify the renewable penetration ratio. Furthermore, Ahmad et al. [20] clarified the chances and challenges of artificial intelligence (AI) in the new digital energy market, which include safety, privacy, and information security. They concluded that the adoption of AI will promote cheap and clean energy with low costs for sustainability transformation. The IoT-based sustainability can improve efficiency and quality of life [21]. Sovacool et al. [22] studied the lifecycle design of data centers and analyzed cooling/electrical energy consumption. Digitalization of energy systems can contribute to sustainable development goals [23], with improved energy efficiency and clean and secure energy supply.

Fig. 11.1 demonstrates the EI and smart grid, with advanced techniques for analyzing, judging, optimizing, and decision-making. Instead of isolation between each building, a smart grid can achieve a dynamic balance and interaction between different end users, such as buildings, industries, power suppliers, energy traders, and system service suppliers. On the other hand, cloud computing enables data processing and performance prediction for large-scale buildings. Big data and AI can enable system optimization and decision-making.

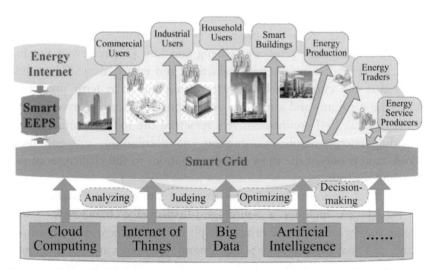

Figure 11.1 Optimization procedure, optimization algorithm, and performance assessment (e.g., time duration and reliability of optimal results).

11.3 Application of Internet of Energy for smart grids

The IoT can provide distributed monitoring services for the smooth running of the smart grid. However, a lack of security, decentralization, transparency, and trust may lead to the application of blockchain-IoT-based smart grid monitoring [24]. Dileep [25] comprehensively reviewed smart grid technologies with smart metering, sensors, and electronic devices. The machine learning technique plays an essential role in multienergy integrations (e.g., thermal and electrical power and fuel) [26,27]. Reka and Dragicevic [28] holistically reviewed the IoT-based smart grid, in terms of energy management and an innovative approach for applications. To achieve a sustainable city, Renugadevi et al. [29] highlighted the IoT-based smart energy grid, with energy management and storage. Wu et al. [1] reviewed sustainable EI with the transition from a things-based to a services-based network. Based on data monitoring and uploading, Hashmi et al. [27] demonstrated demand profiles through the IoT, cloud, and/or fog computing for demand-side management in smart grids. Due to the advantages of high transfer speed, reliability, and robust security, Hui et al. [30] successfully applied 5G in demand response for the development of smart grids. The review provides advanced planning and future directions for 5G applications in smart grids. Bayindir et al. [31] reviewed data transmission, distribution, and energy efficiency of smart grids, providing driving forces to transform from classical grids to smart grids. Lawrence et al. [32] holistically summarized ten questions on building–grid interactions, from the perspectives of technology, communication systems, and standard protocols. Cheng and Lim [33] reviewed the IoT for demand prediction, energy interaction, and energy usage. They indicated that the current technology is mainly on short-term prediction, while the medium- and long-term predictions need further investigations. Motlagh et al. [34] comprehensively reviewed the IoT in energy sectors, including energy supply, transmission and distribution, and end users. Blockchain is one of the most effective solutions to the challenges of privacy and security.

Fig. 11.2 demonstrates EI in smart grids. The structural configuration in Fig. 11.2A includes a source-grid-load-storage framework. Clean energy supply units include renewable plants, nuclear power, and pumped hydro and coal-based power plants. Energy consumers include buildings, transportations, and industries. As intermediate energy storages, electrochemical batteries serve to dynamically balance the power supply and

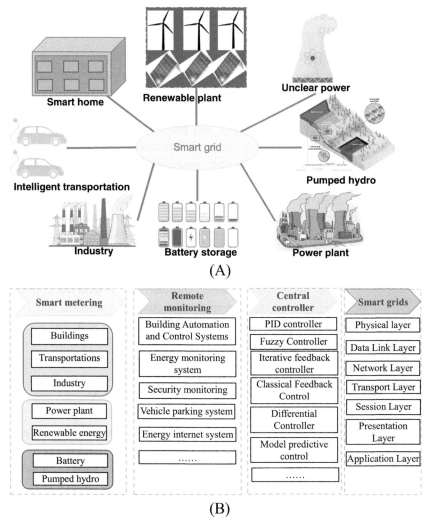

Figure 11.2 Internet of Energy for smart grids: (A) structural configuration; (B) IoT techniques.

energy demand. Fig. 11.2B demonstrates the critical steps for IoT techniques, including smart metering for data collection, remote monitoring for automation, control and security, central controller, and smart grids. Smart grids include several layers, that is, physical layer, data-link layer, network layer, transport layer, session and presentation layer, application layer, etc.

11.4 Digitalized energy network for sustainability transformation

The mutual relationship between sustainability and digitalization is complex. On the one hand, digitalization in internet data centers with considerable power consumption will impose great challenges on sustainability [35], On the other hand, digitalization with smart metering and sensors can optimize energy use behaviors and promote model predictive controls. Specifically, digitalization (e.g., information and communication technologies) will lead to production, usage, and disposal issues, while at the same time improving energy efficiency [36].

Generally, digitalization in sustainability transformation mainly includes energy use behaviors [37], weather prediction, and climate change adaptation [38,39]. Balogun et al. [38] evaluated the digitalization potentials in early warning and emergency response, and climate change adaptation measures. Williams et al. [40] comprehensively reviewed the impact of 5G applications on occupants' behavior and energy-use patterns, together with the embodied energy use when constructing 5G. Husaini and Lean [37] indicated that digitalization is highly correlated with energy-use behavior, showing promising prospects in energy efficiency, sustainability, and energy consumption. Based on the sustainable development goals of the UN, Mondejar et al. [41] acknowledge the necessary elements for a sustainable planet, which include digitalization, IoT, and AI.

Based on smart sensors and digital simulation, Habibi et al. [42] studied the effects of digitalization on user behavior for energy efficiency. Results showed that dynamic interaction between users and control systems through digitalization is highly necessary. Silvestre et al. [43] explored decarbonization, digitalization, and decentralization for energy system transformation. Results showed that the digital revolution will lead to a significant decarbonization potential in the electricity sector. Loock et al. [44] indicated that digitalization can promote the development of business models, contributing to a sustainable energy transition. Koirala et al. [45] explored community energy storages for sustainable transitions with coordination and interaction among different components. Gao et al. [46] studied the relationship between digitization and green total factor energy efficiency (GTFEE) based on 213 Chinese cities. Results showed that the ICT can reduce the GTFEE, and the upper threshold is identified by using the dynamic threshold model. To achieve sustainable development goals under

COVID-19, Çelik et al. [23] explored how digitalization technologies (e.g., smart grid, blockchain, and the IoT) can contribute to global net-zero emissions. Based on the results, it can be concluded that digitalization will play a significant role in securing power supply from cyberattacks and climate change, and as a frontier guideline for politicians and legislators.

Fig. 11.3 demonstrates a 5G architecture for sustainability transformation, consisting of a connection level, conversion level, cyber level, cognition level, and configuration level.

Wu et al. [48] proposed a blockchain-based transactive energy framework for the decarbonization of energy districts, consisting of buildings, EVs, and occupants. Results indicated that the blockchain technique can contribute to the decentralization and transparency of transactive energy. Fig. 11.4 demonstrates an IoT-based transactive energy framework with renewable supplies and energy demands. Renewable suppliers will trade with the whole energy market, while end users will communicate with the retail energy market. Equal opportunities can be provided to all participants through energy transactions between wholesale and retail markets [49].

Fig. 11.4 demonstrates a transactive energy framework with the IoT. Data can be collected through services from the supplier and consumer sides. Through bidirectional interactions between the service layer and the middle layer with big data, cloud of things, and AI, the data will go

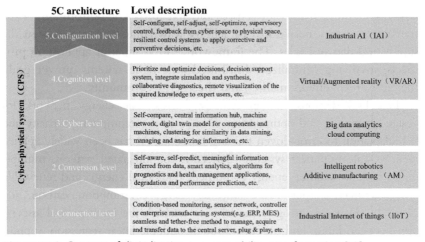

Figure 11.3 Concept of digitalization in sustainability transformation [47].

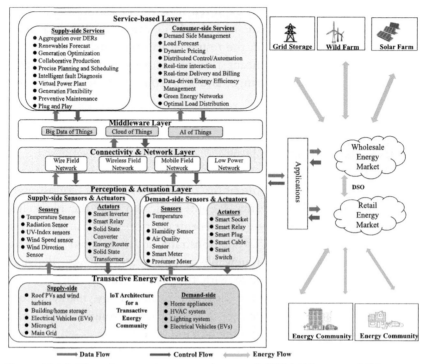

Figure 11.4 IoT-based transactive energy framework [48]. *IoT*, Internet of Things.

through the connectivity and network layers. The network layer will have bidirectional interactions with the perception and actuation layer, which includes demand-side sensors and actuators. The bottom line of the transactive energy network is to achieve a dynamic balance between the supply and demand sides. The framework shows superior application prospects, involving centralized/distributed renewable systems, power grids, the entire retail energy market, and the energy community.

Fig. 11.5 demonstrates an AI-assisted cross-scale energy network for design, operation, and optimization. Due to its supercomputing capacity and fast adaptation to error-driven training processes, AI has been widely applied in cross-scale nonlinear systems. Generally, for integrated systems, multiple inputs include weather, energy consumers, traditional energy, and renewable energy sources; outputs mainly include energy consumption, carbon, and cost. The AI-based surrogate model is to establish the mathematical association between inputs and outputs, through multiple linear regression, support vector regression, and backpropagation. Subsequently, the well-trained

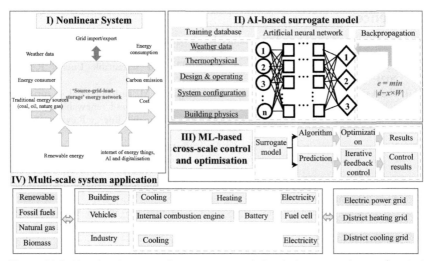

Figure 11.5 AI-assisted cross-scale energy network for design, operation, and optimization. *AI*, Artificial intelligence.

surrogate model can be applied for optimization through algorithms and advanced controls through fast prediction. Furthermore, with respect to the multiscale system application, bidirectional interactions among energy sources, end users, and hybrid grids can be achieved, to guarantee energy supply reliability and a clean transition.

11.5 Prospects of Internet of Energy and digitalization

Future prospects of the IoE and digitalization are summarized below:

- Edge computing can help address the digitization of smart grids and promote sustainability [50], with respect to state, signal, event, and customer analysis.
- The compatibility between IoT and renewable energy requires advanced technologies and regulatory frameworks from policymakers, regulatory bodies, and practitioners [51].
- Trade-off strategies are required for sustainability between improved energy efficiency and production/usage/disposal issues, due to digitalization.
- Energy digitalization will increase machine learning for feedback control, automation, optimization, and the trade-off between cloud and edge computing [52].

References

[1] Wu Y, Wu Y, Guerrero JM, Vasquez JC. A comprehensive overview of framework for developing sustainable energy internet: from things-based energy network to services-based management system. Renewable and Sustainable Energy Reviews 2021;150:111409.

[2] Shaukat N, Ali SM, Mehmood CA, Khan B, Jawad M, Farid U, et al. A survey on consumers empowerment, communication technologies, and renewable generation penetration within Smart Grid. Renewable and Sustainable Energy Reviews 2018;81 (1):1453−75.

[3] Berjawi AEH, Walker SL, Patsios C. An evaluation framework for future integrated energy systems: a whole energy systems approach. Renewable and Sustainable Energy Reviews 2021;145:111163.

[4] Ghobakhloo M, Fathi M. Industry 4.0 and opportunities for energy sustainability. Journal of Cleaner Production 2021;295:126427.

[5] Liu L, Guo X, Lee C. Promoting smart cities into the 5G era with multi-field Internet of Things (IoT) applications powered with advanced mechanical energy harvesters. Nano Energy 2021;88:106304.

[6] Inderwildi O, Zhang C, Wang X, Kraft M. The impact of intelligent cyber-physical systems on the decarbonization of energy. Energy & Environmental Science 2020;13:744−71.

[7] Fouquet R, Hippe R. Twin transitions of decarbonisation and digitalisation: a historical perspective on energy and information in European economies. Energy Research & Social Science 2022;91:102736.

[8] Zhou Y, Zheng SQ. Machine-learning based hybrid demand-side controller for high-rise office buildings with high energy flexibilities. Applied Energy 2020;262:114416.

[9] Liu Z, Xiang Z, Ying S, Yuekuan Z. Advanced controls on energy reliability, flexibility and occupant-centric control for smart and energy-efficient buildings. Energy and Buildings 2023;297:113436.

[10] Wu H, Xue Y, Hao Y, Ren S. How does internet development affect energy-saving and emission reduction? Evidence from China. Energy Economics 2021;103:105577.

[11] Haldar A, Sethi N. Environmental effects of Information and Communication Technology—exploring the roles of renewable energy, innovation, trade and financial development. Renewable and Sustainable Energy Reviews 2022;153:111754.

[12] Zhou Y, Siqian Z, Jiachen L, Yunlong Z. A cross-scale modelling and decarbonisation quantification approach for navigating Carbon Neutrality Pathways in China. Energy Conversion and Management 2023;297:117733.

[13] Sareen S. Digitalisation and social inclusion in multi-scalar smart energy transitions. Energy Research & Social Science 2021;81:102251.

[14] Europe's Digital Decade: digital targets for 2030. <https://ec.europa.eu/info/strategy/priorities-2019-2024/europe-fit-digital-age/europes-digital-decade-digital-targets-2030_en>.

[15] Ren S, Hao Y, Xu L, Wu H, Ba N. Digitalization and energy: how does internet development affect China's energy consumption? Energy Economics 2021;98:105220.

[16] Hao Y, Li Y, Guo Y, Chai J, Yang C, Wu H. Digitalization and electricity consumption: does internet development contribute to the reduction in electricity intensity in China? Energy Policy 2022;164:112912.

[17] Zhang H, Gao S, Zhou P. Role of digitalization in energy storage technological innovation: evidence from China. Renewable and Sustainable Energy Reviews 2023;171:113014.

[18] Wu Y, Wu Y, Guerrero JM, Vasquez JC. Digitalization and decentralization driving transactive energy Internet: key technologies and infrastructures. International Journal of Electrical Power & Energy Systems 2021;126:106593.
[19] Tayal D. Achieving high renewable energy penetration in Western Australia using data digitisation and machine learning. Renewable and Sustainable Energy Reviews 2017;80:1537−43.
[20] Ahmad T, Zhang D, Huang C, Zhang H, Dai N, Song Y, et al. Artificial intelligence in sustainable energy industry: status quo, challenges and opportunities. Journal of Cleaner Production 2021;289:125834.
[21] Nižetić S, Šolić P, González-de DLI, Patrono L. Internet of Things (IoT): opportunities, issues and challenges towards a smart and sustainable future. Journal of Cleaner Production 2020;274:122877.
[22] Sovacool BK, Monyei CG, Upham P. Making the internet globally sustainable: technical and policy options for improved energy management, governance and community acceptance of Nordic datacenters. Renewable and Sustainable Energy Reviews 2022;154:111793.
[23] Çelik D, Meral ME, Waseem M. Investigation and analysis of effective approaches, opportunities, bottlenecks and future potential capabilities for digitalization of energy systems and sustainable development goals. Electric Power Systems Research 2022;211:108251.
[24] Baidya S, Potdar V, Ray PP, Nandi C. Reviewing the opportunities, challenges, and future directions for the digitalization of energy. Energy Research & Social Science 2021;81:102243.
[25] Dileep G. A survey on smart grid technologies and applications. Renewable Energy 2020;146:2589−625.
[26] Ahmad T, Madonski R, Zhang D, Huang C, Mujeeb A. Data-driven probabilistic machine learning in sustainable smart energy/smart energy systems: key developments, challenges, and future research opportunities in the context of smart grid paradigm. Renewable and Sustainable Energy Reviews 2022;160:112128.
[27] Hashmi SA, Ali CF, Zafar S. Internet of things and cloud computing-based energy management system for demand side management in smart grid. International Journal of Energy Research 2020;45(1):1007−22.
[28] Reka SS, Dragicevic T. Future effectual role of energy delivery: a comprehensive review of Internet of Things and smart grid. Renewable and Sustainable Energy Reviews 2018;91:90−108.
[29] Renugadevi N, Saravanan S, Sudha CMN. IoT based smart energy grid for sustainable cites. Materials Today: Proceedings 2021;. Available from: https://doi.org/10.1016/j.matpr.2021.02.270.
[30] Hui H, Ding Y, Shi Q, Li F, Song Y, Yan J. 5G network-based Internet of Things for demand response in smart grid: a survey on application potential. Applied Energy 2020;257:113972.
[31] Bayindir R, Colak I, Fulli G, Demirtas K. Smart grid technologies and applications. Renewable and Sustainable Energy Reviews 2016;66:499−516.
[32] Lawrence TM, Boudreau MC, Helsen L, Henze G, et al. Ten questions concerning integrating smart buildings into the smart grid. Building and Environment 2016;108:273−83.
[33] Cheng YL, Lim MH. Impact of Internet of Things Paradigm towards energy consumption prediction: a systematic literature review. Sustainable Cities and Society 2022;78:103624.
[34] Hossein Motlagh N, Mohammadrezaei M, Hunt J, Zakeri B. Internet of Things (IoT) and the energy sector. Energies 2020;13(2):494.

[35] Kez DA, Foley AM, Laverty D, Rio DFD, Sovacool B. Exploring the sustainability challenges facing digitalization and internet data centers. Journal of Cleaner Production 2022;371:133633.

[36] Lange S, Pohl J, Santarius T. Digitalization and energy consumption. Does ICT reduce energy demand? Ecological Economics 2020;176:106760.

[37] Husaini DH, Hooi Lean H. Digitalization and energy sustainability in ASEAN. Resources, Conservation and Recycling 2022;184:106377.

[38] Balogun AL, Marks D, Sharma R, Shekhar H, Balmes C, Maheng D, et al. Assessing the potentials of digitalization as a tool for climate change adaptation and sustainable development in urban centres. Sustainable Cities and Society 2020;53:101888.

[39] Zhou Y. Climate change adaptation with energy resilience in energy districts—A state-of-the-art review. Energy and Buildings 2023;279:112649.

[40] Williams L, Sovacool BK, Foxon TJ. The energy use implications of 5G: Reviewing whole network operational energy, embodied energy, and indirect effects. Renewable and Sustainable Energy Reviews 2022;157:112033.

[41] Mondejar ME, Avtar R, Diaz HLB, Dubey RK, et al. Digitalization to achieve sustainable development goals: steps towards a Smart Green Planet. Science of The Total Environment 2021;794:148539.

[42] Habibi S. Micro-climatization and real-time digitalization effects on energy efficiency based on user behavior. Building and Environment 2017;114:410−28.

[43] Silvestre MLD, Favuzza S, Sanseverino ER, Zizzo G. How decarbonization, digitalization and decentralization are changing key power infrastructures. Renewable and Sustainable Energy Reviews 2018;93:483−98.

[44] Loock M. Unlocking the value of digitalization for the European energy transition: a typology of innovative business models. Energy Research & Social Science 2020;69:101740.

[45] Koirala BP, van Oost E, van der Windt H. Community energy storage: a responsible innovation towards a sustainable energy system? Applied Energy 2018;231:570−85.

[46] Gao D, Li G, Yu J. Does digitization improve green total factor energy efficiency? Evidence from Chinese 213 cities. Energy 2022;247:123395.

[47] Chen X, Despeisse M, Johansson B. Environmental sustainability of digitalization in manufacturing: a review. Sustainability 2020;12(24):10298.

[48] Wu Y, Wu Y, Guerrero JM, Vasquez JC. Decentralized transactive energy community in edge grid with positive buildings and interactive electric vehicles. International Journal of Electrical Power & Energy Systems 2022;135:107510.

[49] Chong PHJ, Seet B-C, Chai M, Rehman SU. Smart grid and innovative frontiers in telecommunications. Smart GIFT 2018. Available from: https://doi.org/10.1007/978-3-319-94965-9.

[50] Feng C, Wang Y, Chen Q, Ding Y, Strbac G, Kang C. Smart grid encounters edge computing: opportunities and applications. Advances in Applied Energy 2021;1:100006.

[51] Mishra R, Naik BKR, Rakesh DR, Kumar M. Internet of Things (IoT) adoption challenges in renewable energy: a case study from a developing economy. Journal of Cleaner Production 2022;371:133595.

[52] Isaksson AJ, Harjunkoski I, Sand G. The impact of digitalization on the future of control and operations. Computers & Chemical Engineering 2018;114:122−9.

CHAPTER 12

Application of big data and cloud computing for the development of integrated smart building transportation energy systems

Yuekuan Zhou[1,2,3,4]
[1]Sustainable Energy and Environment Thrust, Function Hub, The Hong Kong University of Science and Technology (Guangzhou), Nansha, Guangdong, P.R. China
[2]Department of Mechanical and Aerospace Engineering, The Hong Kong University of Science and Technology, Clear Water Bay, Hong Kong SAR, P.R. China
[3]HKUST Shenzhen-Hong Kong Collaborative Innovation Research Institute, Futian, Shenzhen, P.R. China
[4]Division of Emerging Interdisciplinary Areas, The Hong Kong University of Science and Technology, Clear Water Bay, Hong Kong SAR, P.R. China

12.1 Introduction

Electrification and hydrogenation can contribute to carbon-neutral transformation [1] with high energy density and a fast power response. Digitalization and decentralization are mainstreams of the future transactive Energy Internet [2]. The deployment of distributed renewable systems will lead to the role transition from traditional consumers to prosumers. The prosumer-centric energy-sharing system requires different management strategies [3], considering dynamic grid prices, stochastic demands, thermal/electrical energy storages, etc. The integrated energy sharing between prosumer-based buildings and various types of vehicles [4] can mitigate grid power fluctuations, reduce operating costs, and enhance renewable self-consumption. However, the associated aging of intermediate storages for energy sharing (such as electrochemical batteries) proposes new challenges to the techno-economic feasibility of the energy-sharing framework.

With the rapid development of smart metering and sensor technologies, big data can be collected in integrated building transportation energy systems. Based on the big database, artificial intelligence can enable the integrated system to become smart with considerable energy-saving potential. Zhu et al. [5] comprehensively reviewed data-driven techniques for load forecasting. Furthermore, high computation is required considering both epistemic and aleatory uncertainties in multiple input parameters. Zhou et al. [6] developed

Advances in Digitalization and Machine Learning for Integrated Building-Transportation Energy Systems
DOI: https://doi.org/10.1016/B978-0-443-13177-6.00005-9

a surrogate model with machine learning (ML) to predict the onsite power supply of a phase change material (PCM)–photovoltaic (PV) system. Uncertainty-based optimization can effectively improve equivalent overall output energy in different climate zones [7]. Guo et al. [8] comprehensively reviewed cloud computing in smart energy systems. Tian et al. [9] applied cloud computing for confidentiality preservation in integrated multienergy systems. Mostafa et al. [10] studied ML-based renewable dispatch strategies and concluded that ML was critical for grid power stability and prediction accuracy.

The integrated smart building transportation energy system involves multiagent power/energy interactions. Zhou [11] studied an ML-based smart community, concerning capacity sizing, dynamic scheduling, and power dispatch. Thomas et al. [12] studied power dispatching and scheduling strategies for cost savings in an electric vehicle (EV)-integrated PV-building system. Stochastic scheduling is beneficial for daily cost saving.

In this chapter, new energy typology transformations are provided for flexible unit integrations, multidirectional interactions, and techno-economic analysis for multistakeholders. Modeling techniques are comprehensively reviewed for power-signal-cash flows and energy-information-trading. In terms of frequent data exchange and power trading in digital multienergy systems, fundamental roles and functions of cloud computing are provided, in terms of data privacy and transformation security, data visualization, fault detection and diagnosis, automation, and so on. This chapter also deals with the challenges and future prospects of big data and cloud computing for sustainable and intelligent multienergy systems.

12.2 New energy typologies in decentralized power suppliers/consumers

With different energy forms for end users and the fast development of electrification in integrated energy systems, new energy typologies are gradually replacing traditional fossil fuel-supported energy systems. Table 12.1 summarizes new energy typologies for sustainable transition. Kim et al. [19] comprehensively reviewed an integrated smart home and city framework, including monitoring, diagnostics and controlling, intelligent energy management, and a city-level based energy network. For the carbon neutrality transition, Liu et al. [20] comprehensively reviewed AI applications in large-scale renewable systems, in terms of power supply, distribution, and demand side. Zhou [21] formulated a spatiotemporal energy network for regional energy

Table 12.1 New energy typologies for sustainability.

System typologies	Study	Methods	Results
Sustainable airport-land ecosystem	Zhou [13]	Frontier energy framework	Sustainability transformation in land and air transportations through electrification and hydrogenation
Building-vehicle sharing	Zhou and Cao [14]	Flexibility assessment in simulation platform	EV integration can enhance energy flexibility in grid power shifting.
Building-vehicle-building sharing	Zhou et al. [15]	Multiobjective optimization	Optimal solutions show lowest carbon emissions, lower costs, but highest energy flexibility.
Cross-regional buildings-vehicles-buildings network	Zhou [16]	Comparative analysis	Guidelines on energy pricing policy and energy interaction modes.
Cross-regional energy sharing network in Bay Area	Zhou [17]	Dynamic trading pricing	Stakeholders are incentivized to participate in the formulated energy network under dynamic pricing.
Hydrogen economy-based inter-city energy network	He et al. [18]	Grid power and energy prices	Critical renewable export price, H2 price and internal trading price

EV, Electric vehicle.

sharing and trading, contributing to regional energy balance. An inter-city energy network with hydrogen vehicles can achieve a more even distribution of renewable energy [22], whereas the degradation of fuel cells requires coordination and tradeoff strategies. Results showed that the proposed network is economically feasible only when the renewable export price for H_2 production is at 0.07 $/kWh, the onsite-renewable-generated H_2 is lower than

6.5 $/kg and vehicle-to-building (V2B) electricity is lower than 0.3 $/kWh. An interactive energy sharing community with e-mobility has been proposed by Zhou [23], together with a regression learner-based approach for battery cycling aging prediction. Results showed that the ML approach can help predict and decelerate the battery cycling aging in a multienergy community.

Considering the mismatch between energy supply resources and end-users' demand (i.e., abundant solar-wind energy resources in suburbs and high demand density in city centers), Zhou [13] comprehensively reviewed a renewable-grid-storage-flexibility energy ecosystem for sustainable transitions. Multidiversified energy resources (e.g., geothermal, natural gas, solar-wind, ocean energy, etc.), and mobile energy transmissions (e.g., EVs and hydrogen vehicles) can be exploited for energy flexibility provision. Zheng et al. [24] designed distributed battery storages and studied techno-economic performance with synergistic energy sharing. Renewable self-consumption can be improved, while the net cost can be reduced. Zheng et al. [25] studied prosumer-based peer-to-peer energy sharing under demand-side flexibility. Results indicate an improvement in self-sufficiency of 22.8% through the proposed strategy.

Fig. 12.1 demonstrates a cross-border energy network with fuel cell EVs (FCEVs), renewable systems, power plants, and building districts [13]. Different energy forms are involved, that is, thermal, power, and hydrogen. Renewable energy (i.e., solar, wind, and biomass) from rural or suburban areas can be transmitted into the city center through pipelines or power transmission lines. Mobile FCEVs can carry energy fuels from renewable-abundant regions (i.e., PVs or wind farms) to demand-shortage regions (e.g., city centers or building communities). In addition, countryside biomass as a renewable energy source can be applied to either local energy supply or district heating and power grids. Both thermal and electrical grids are designed to provide fast demand response, power supply, and dynamic supply-demand balance. Backup systems are designed to ensure energy reliability, including centralized combined heat and power with carbon capture and storage, boilers, heat pumps, and heat recovery.

12.3 Modeling development in advanced energy-information-trading networks and power-signal-cash flows

Fig. 12.2 demonstrates an electrification-based sharing network with power-cash-information flows. Contrary to the power flow direction, cash flows can be noticed between different stakeholders. However, bilateral information

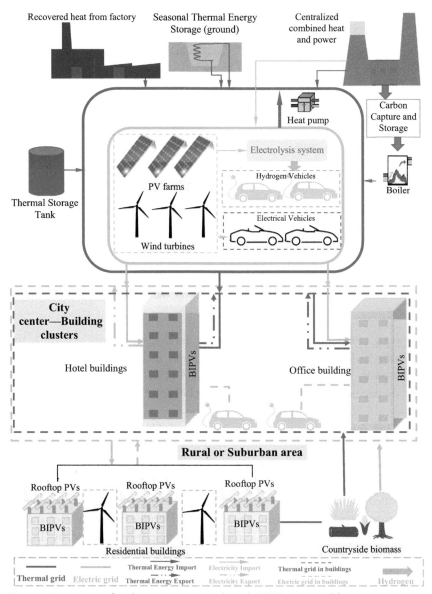

Figure 12.1 A cross-border energy network with FCEVs, renewable systems, power plants, and building districts [13].

communication will happen for real-time monitoring, dynamic response, and controls. To investigate the techno-economic performance of integrated building-transportation systems, cross-scale modeling techniques are required,

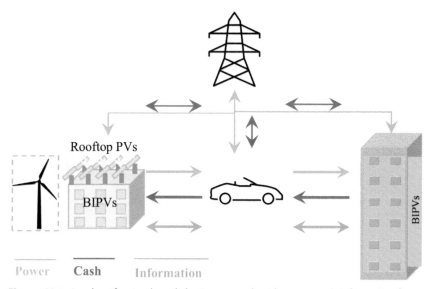

Figure 12.2 An electrification-based sharing network with power-cash-information flows.

involving components, energy conversion devices, subsystems, and entire systems. In academia, modeling techniques mainly include user-defined components with multisoftware coupling (MATLAB/Simulink models) and already-developed components (e.g., TRNSYS, HOMER, etc.) from the library. Table 12.2 provides state-of-the-art modeling development of advanced energy-sharing networks. Zhou et al. [33] investigated the techno-economic performance of a district energy-sharing network, with the transition from a negative toward a positive paradigm. Results indicate an increase in net present value from -7.182×10^7 to 5.164×10^8 HK\$. Liu et al. [34] developed an energy community platform for peer-to-peer (P2P) energy sharing and trading. Based on the developed platform with a peer trading strategy, net grid imports and carbon emissions can be reduced by 18.54% and 1594.13 tons, respectively. Table 12.2 shows the modeling development of advanced energy-sharing networks regarding system configuration, method, objective, and research results. Prencipe et al. [26] developed a mathematical programming model on vehicle-to-grid (V2G) interaction and concluded that the daily trip cost of vehicles can be completely covered. An interactive zero energy-sharing network by Zhou [32] with P2P energy sharing and V2B/V2G interactions can improve the off-peak renewable energy shifting ratio, decelerate the battery relative capacity, and achieve cost savings. Chen et al. [35] developed an energy platform for P2P energy trading.

Table 12.2 Modeling development of advanced energy-sharing networks.

Systems	Study	Methods	Objectives	Results
Vehicle-to-grid	Prencipe et al. [26]	A mathematical programming model	Vehicle daily charging costs	Vehicle-to-grid operations can completely cover daily trip costs by selling power to grid during peak load periods.
Renewables–buildings–vehicles energy-sharing network	Zhou et al. [27]	TRNSYS model	Battery degradation and operating costs	Hierarchical control strategy can slow down battery aging and reduce operating costs.
Static battery-EVs-P2P energy-sharing framework	Lyu et al. [28]	A decentralized algorithm for a peer-to-peer transactive framework	Agent information privacy protection	High privacy protection with economic welfare for smart buildings
$V2B^2$ layouts	Barone et al. [29]	An energy management scheme with $V2B^2$ layouts	Grid interaction and net present value	13.6%—71.2% drop in grid interaction with net present value at 18.0 k€
	Buonomano [30]	Modeling and simulation	Fossil fuel consumption and economic performance	Renewable energy can be fully utilized to gradually replace fossil fuel consumption
A vehicle-to-microgrid system with buildings	Wang et al. [31]	Collaborations between lithium-ion batteries and fuel cells	Battery degradation and CO_2 emissions	CO_2 emissions can be reduced by 515.56 tons in the formulated novel framework.
An interactive zero energy-sharing network	Zhou [32]	Simulation platform development	Equivalent relative capacity, grid import costs, and CO_2 emissions	The developed EMS can improve off-peak renewable energy shifting ratio, decelerate the battery relative capacity, and achieve cost savings.

The platform can flexibly integrate surrogate models with advanced ML algorithms and optimization approaches for energy trading. Barone et al. [36] developed a model to simulate the building to V2B for the zero-energy paradigm transition. The model can accurately quantify grid power interactions and energy consumption.

Concerning multivariable and multiobjective optimization, Fernandez et al. [37] developed a bi-level optimization platform for energy sharing and trading. By operating the developed platform, the grid pricing scheme is the main driving force for revenue sharing. Bigdeli [38] optimized a PV/fuel cell/battery power system through combined optimization algorithms in load sharing. Results showed that a fuzzy logic controller with quantum-behaved particle swarm optimization performed the best. Elmorshedy et al. [39] optimized the system design and operation of an isolated system with renewable batteries and converters. Optimal capacities and energy cost analyses of each component are identified.

12.4 Big data, cloud computing, and digital twin in interactive multienergy networks

The further development of the Energy Internet generates big data from smart metering/sensors [40]. Wireline and wireless communication technologies are employed throughout the generation, transmission, distribution, and end-user side. Cloud computing can contribute to smart energy management, demand-side management, and power dispatching in energy hubs [41]. In addressing the sensor-collected data transformation failure, Jiang [42] developed a Markov chain-based data aggregation algorithm for data sharing and information exchange. Stergiou et al. [43] applied cloud computing for data privacy and transformation security, which can also be applied in building and grid automation, and vehicle-to-vehicle communication. Zhang et al. [44] studied cloud computing-based global optimization on both battery aging and energy consumption for vehicles. The vehicle-in-the-loop simulation platform with distributed EMS outperforms the rule-based EMS with a cost saving of 6.8%. Zhou et al. [45] comprehensively reviewed smart energy management with big data from the perspective of power supply, microgrids, and collaborative operations. Zhou and Yang [46] studied big data-based energy-use behavior for forecasting energy consumption and energy saving. Wu et al. [47] comprehensively reviewed energy-based, communication-based, and service-based Energy Internet with digitalization and cloud computing. Cutting-edge technology can help promote a sustainable transition.

Keirstead et al. [48] comprehensively reviewed modeling techniques on urban energy systems, including buildings and transportations, using cloud computing. The review highlights the necessity for multidisciplinary work on multienergy systems in smart cities. Tuballa and Abundo [49] systematically reviewed the development of smart grid technologies, from the perspective of features, functionalities, and characteristics. The development of the smart grid indicates multidisciplinary collaboration, aspirations, and shared lessons. Wang et al. [50] comprehensively reviewed applications of digital twin technologies in energy management, including system design, operation, and prediction. The dynamic interaction between the digital model and the real prototype can help improve system performance and optimize the operation. Ahmad and Zhang [51] applied IoT to smart energy systems, in terms of detection, data mining, and visualization. The techniques applied include cloud computing, edge computing, and quantum computing to provide supercomputing services.

Fig. 12.3 demonstrates digital twin-based construction and operation in building sectors. Regarding the BIM platform for building digitization, as shown in Fig. 12.3A, the entire process involves model development, design, management, and onsite testing. The real-time monitoring and observation, as shown in Fig. 12.3B, include precise mapping, virtual-reality interaction, software definition, and intelligent intervention. Both data-driven and model-driven digital twin techniques are available. The data-driven digital twin technique highly relies on the amount and quality of the data, while the model-driven digital twin technique relies on the mathematical models following physical mechanisms.

Fig. 12.4 demonstrates digital twin models for semivirtual building prototype interaction and ML-based models in parameter optimization. As demonstrated in Fig. 12.4A, digital twin involves precise mapping, virtual reality interaction, software definition, and intelligent intervention. On the one hand, the onsite experimental platform was established. On the other hand, a twin-simulated building is established to achieve energy, information, and cost interactions. Compared with the onsite experimental platform, the twin model is cost-effective and time-efficient. Outputs of the twin model include cooling/heating/electric loads, energy consumption, peak power, economic cost, and carbon emissions. By feeding back the outputs into the building prototype, semivirtual building prototype interaction can be achieved.

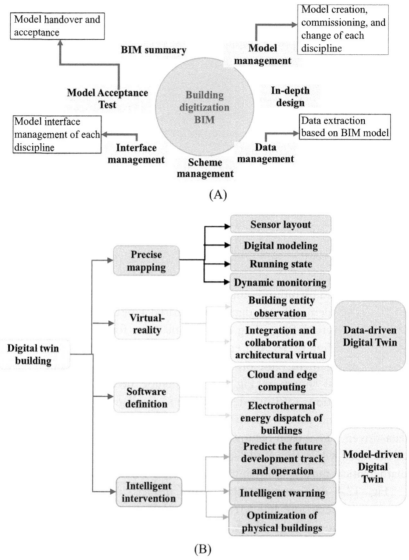

Figure 12.3 Digital twin technologies: (A) Building Information Model (BIM) for construction; (B) dynamic monitoring.

In terms of multiple parameter optimization, as shown in Fig. 12.4B, input parameters across the material–component–building–districts are variables, and the output objectives include cooling/heating/electric loads, energy consumption, peak power, economic performance, and carbon

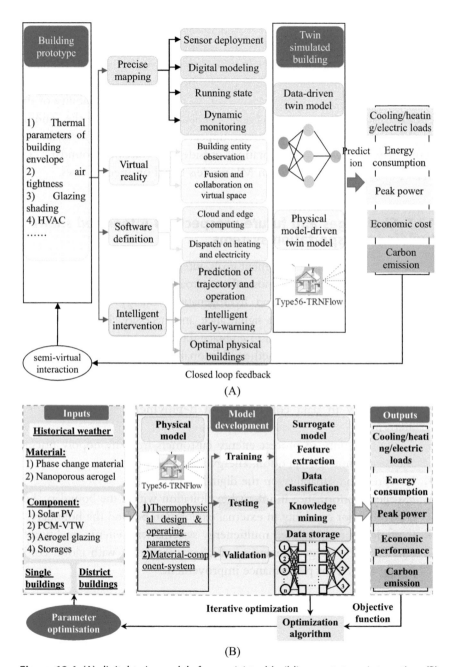

Figure 12.4 (A) digital twin models for semivirtual building prototype interaction; (B) machine learning-based model for parameter optimization.

emissions. The aim of the ML-based model is to establish a straightforward association between inputs and outputs. Through iterative calculations, optimal variables can be identified to achieve the best performance. The application of ML can speed up the optimization process, due to its supercomputing performance. However, to guarantee the reliability of the optimal results, several critical points need to be noticed: (1) the training database needs to include a lower/upper threshold of each variable to avoid new knowledge exploration outside the variable boundary; (2) hyperparameter optimization in ML models based on specific cases.

12.5 Challenges and future prospects of integrated smart building transportation systems

To achieve sustainable development goals, integrated smart building transportation systems are one of the most effective roadmaps, especially with renewable energy, electrification, and hydrogenation. Energy digitalization and cloud computing are of great significance and necessity. However, several challenges need to be identified to embrace the future energy transformation.

1. The open question on whether additional energy consumption in telecommuting exceeds energy savings requires further investigation [52], due to the inadequate database.
2. Uncertainty in data collection and deployment in smart sensors/ metering.
3. A statutory duty to protect energy customers from monopoly pricing and to promote competition, energy supply security.
4. Interpretation capability on the digital-twin model.
5. Excellent performance in internal exploitation within the boundary, but relatively poor capability in external exploration beyond the boundary.
6. Complexity in integrated multienergy systems with energy quality differences and energy conversion efficiency, together with cloud computing and AI for performance improvement.

References

[1] Zhou Y. Transition towards carbon-neutral districts based on storage techniques and spatiotemporal energy sharing with electrification and hydrogenation. Renewable and Sustainable Energy Reviews 2022;162:112444.
[2] Wu Y, Wu Y, Guerrero JM, Vasquez JC. Digitalization and decentralization driving transactive energy Internet: key technologies and infrastructures. International Journal of Electrical Power & Energy Systems 2021;126:106593.

[3] Zafar R, Mahmood A, Razzaq S, Ali W, Naeem U, Shehzad K. Prosumer based energy management and sharing in smart grid. Renewable and Sustainable Energy Reviews 2018;82:1675−84.

[4] Zhou Y, Cao S, Hensen JLM, Lund PD. Energy integration and interaction between buildings and vehicles: a state-of-the-art review. Renewable and Sustainable Energy Reviews 2019;114:109337.

[5] Zhu J, Dong H, Zheng W, Li S, Huang Y, Xi L. Review and prospect of data-driven techniques for load forecasting in integrated energy systems. Applied Energy 2022;321:119269.

[6] Zhou Y, Zheng S, Zhang G. Machine-learning based study on the on-site renewable electrical performance of an optimal hybrid PCMs integrated renewable system with high-level parameters' uncertainties. Renewable Energy 2020;151:403−18.

[7] Zhou Y, Zheng S, Zhang G. Machine learning-based optimal design of a phase change material integrated renewable system with on-site PV, radiative cooling and hybrid ventilations—study of modelling and application in five climatic regions. Energy 2020;192:116608.

[8] Guo C, Luo F, Cai Z, Dong ZY. Integrated energy systems of data centers and smart grids: state-of-the-art and future opportunities. Applied Energy 2021;301:117474.

[9] Tian N, Ding T, Yang Y, Guo Q, Sun H, Blaabjerg F. Confidentiality preservation in user-side integrated energy system management for cloud computing. Applied Energy 2018;231:1230−45.

[10] Mostafa N, Ramadan HSM, Elfarouk O. Renewable energy management in smart grids by using big data analytics and machine learning. Machine Learning with Applications 2022;9:100363.

[11] Zhou Y. Advances of machine learning in multi-energy district communities-mechanisms, applications and perspectives. Energy and AI 2022;10:100187.

[12] Thomas D, Deblecker O, Ioakimidis CS. Optimal operation of an energy management system for a grid-connected smart building considering photovoltaics' uncertainty and stochastic electric vehicles' driving schedule. Applied Energy 2018;210:1188−206.

[13] Zhou Y. Low-carbon transition in smart city with sustainable airport energy ecosystems and hydrogen-based renewable-grid-storage-flexibility. Energy Reviews 2022;1 (1):100001.

[14] Zhou Y, Cao S. Energy flexibility investigation of advanced grid-responsive energy control strategies with the static battery and electric vehicles: a case study of a high-rise office building in Hong Kong. Energy Conversion and Management 2019;199:111888.

[15] Zhou Y, Cao S, Kosonen R, Hamdy M. Multi-objective optimisation of an interactive buildings-vehicles energy sharing network with high energy flexibility using the Pareto archive NSGA-II algorithm. Energy Conversion and Management 2020;218:113017.

[16] Zhou Y. Energy sharing and trading on a novel spatiotemporal energy network in Guangdong-Hong Kong-Macao Greater Bay Area. Applied Energy 2022;318:119131.

[17] Zhou Y. Incentivising multi-stakeholders' proactivity and market vitality for spatiotemporal microgrids in Guangzhou-Shenzhen-Hong Kong Bay Area. Applied Energy 2022;328:120196.

[18] He Y, Zhou Y, Liu J, Liu Z, Zhang G. An inter-city energy migration framework for regional energy balance through daily commuting fuel-cell vehicles. Applied Energy 2022;324:119714.

[19] Kim H, Choi H, Kang H, An J, Yeom S, Hong T. A systematic review of the smart energy conservation system: from smart homes to sustainable smart cities. Renewable and Sustainable Energy Reviews 2021;140:110755.

[20] Liu Z, Sun Y, Xing C, Liu J, He Y, Zhou Y, et al. Artificial intelligence powered large-scale renewable integrations in multi-energy systems for carbon neutrality transition: challenges and future perspectives. Energy and AI 2022;10:100195.

[21] Zhou Y. Energy sharing and trading on a novel spatiotemporal energy network in Guangdong-Hong Kong-Macao Greater Bay Area. Applied Energy 2022;318:119131.

[22] He Y, Zhou Y, Liu J, Liu Z, Zhang G. An inter-city energy migration framework for regional energy balance through daily commuting fuel-cell vehicles. Applied Energy 2022;324:119714.

[23] Zhou Y. A regression learner-based approach for battery cycling ageing prediction—advances in energy management strategy and techno-economic analysis. Energy 2022;256:124668.

[24] Zheng S, Huang G, Lai ACK. Techno-economic performance analysis of synergistic energy sharing strategies for grid-connected prosumers with distributed battery storages. Renewable Energy 2021;178:1261−78.

[25] Zheng S, Jin X, Huang G, Lai ACK. Coordination of commercial prosumers with distributed demand-side flexibility in energy sharing and management system. Energy 2022;248:123634.

[26] Prencipe LP, van Essen JT, Caggiani L, Ottomanelli M, Correia GHA. A mathematical programming model for optimal fleet management of electric car-sharing systems with vehicle-to-grid operations. Journal of Cleaner Production 2022;368:133147.

[27] Zhou Y, Cao S, Hensen JLM, Hasan A. Heuristic battery-protective strategy for energy management of an interactive renewables−buildings−vehicles energy sharing network with high energy flexibility. Energy Conversion and Management 2020;214:112891.

[28] Lyu C, Jia Y, Xu Z. Fully decentralized peer-to-peer energy sharing framework for smart buildings with local battery system and aggregated electric vehicles. Applied Energy 2021;299:117243.

[29] Barone G, Buonomano A, Forzano C, Giuzio GF, Palombo A, Russo G. Energy virtual networks based on electric vehicles for sustainable buildings: system modelling for comparative energy and economic analyses. Energy 2022;242:122931.

[30] Buonomano A. Building to vehicle to building concept: a comprehensive parametric and sensitivity analysis for decision making aims. Applied Energy 2020;261:114077.

[31] Wang B, Yu X, Xu H, Wu Q, Wang L, Huang R, et al. Scenario analysis, management, and optimization of a new vehicle-to-micro-grid (V2μG) network based on off-grid renewable building energy systems. Applied Energy 2022;325:119873.

[32] Zhou Y. Energy planning and advanced management strategies for an interactive zero-energy sharing network (buildings and electric vehicles) with high energy flexibility and electrochemical battery cycling aging. Hong Kong Polytechnic University−Dissertations, 2021.

[33] Zhou Y, Cao S, Hensen JLM. An energy paradigm transition framework from negative towards positive district energy sharing networks—battery cycling aging, advanced battery management strategies, flexible vehicles-to-buildings interactions, uncertainty and sensitivity analysis. Applied Energy 2021;288:116606.

[34] Liu J, Yang H, Zhou Y. Peer-to-peer trading optimizations on net-zero energy communities with energy storage of hydrogen and battery vehicles. Applied Energy 2021;302:117578.

[35] Chen K, Lin J, Song Y. Trading strategy optimization for a prosumer in continuous double auction-based peer-to-peer market: a prediction-integration model. Applied Energy 2019;242:1121−33.

[36] Barone G, Buonomano A, Calise F, Forzano C, Palombo A. Building to vehicle to building concept toward a novel zero energy paradigm: modelling and case studies. Renewable and Sustainable Energy Reviews 2019;101:625−48.

[37] Fernandez E, Hossain MJ, Mahmud K, Nizami MSH, Kashif M. A Bi-level optimization-based community energy management system for optimal energy sharing and trading among peers. Journal of Cleaner Production 2021;279:123254.

[38] Bigdeli N. Optimal management of hybrid PV/fuel cell/battery power system: a comparison of optimal hybrid approaches. Renewable and Sustainable Energy Reviews 2015;42:377–93.

[39] Elmorshedy MF, Elkadeem MR, Kotb KM, et al. Optimal design and energy management of an isolated fully renewable energy system integrating batteries and supercapacitors. Energy Conversion and Management 2021;245:114584.

[40] Kabalci Y. A survey on smart metering and smart grid communication. Renewable and Sustainable Energy Reviews 2016;57:302–18.

[41] Allahvirdizadeh Y, Moghaddam MP, Shayanfar H. A survey on cloud computing in energy management of the smart grids. International Transactions on Electrical Energy Systems 2019;29:10. Available from: https://doi.org/10.1002/2050-7038.12094.

[42] Jiang D. The construction of smart city information system based on the Internet of Things and cloud computing. Computer Communications 2020;150:158–66.

[43] Stergiou C, Psannis KE, Gupta BB, Ishibashi Y. Security, privacy & efficiency of sustainable cloud computing for big data & IoT. Sustainable Computing: Informatics and Systems 2018;19:174–84.

[44] Zhang Y, Liu H, Zhang Z, Luo Y, Guo Q, Liao S. Cloud computing-based real-time global optimization of battery aging and energy consumption for plug-in hybrid electric vehicles. Journal of Power Sources 2020;479:229069.

[45] Zhou K, Fu C, Yang S. Big data driven smart energy management: from big data to big insights. Renewable and Sustainable Energy Reviews 2016;56:215–25.

[46] Zhou K, Yang S. Understanding household energy consumption behavior: the contribution of energy big data analytics. Renewable and Sustainable Energy Reviews 2016;56:810–19.

[47] Wu Y, Wu Y, Guerrero JM, Vasquez JC. A comprehensive overview of framework for developing sustainable energy internet: from things-based energy network to services-based management system. Renewable and Sustainable Energy Reviews 2021;150:111409.

[48] Keirstead J, Jennings M, Sivakumar A. A review of urban energy system models: approaches, challenges and opportunities. Renewable and Sustainable Energy Reviews 2012;16(6):3847–66.

[49] Tuballa ML, Abundo ML. A review of the development of Smart Grid technologies. Renewable and Sustainable Energy Reviews 2016;59:710–25.

[50] Wang Y, Kang X, Chen Z. A survey of digital twin techniques in smart manufacturing and management of energy applications. Green Energy and Intelligent Transportation 2022;100014.

[51] Ahmad T, Zhang D. Using the internet of things in smart energy systems and networks. Sustainable Cities and Society 2021;68:102783.

[52] O'Brien W, Aliabadi FY. Does telecommuting save energy? A critical review of quantitative studies and their research methods. Energy and Buildings 2020;225:110298.

CHAPTER 13

Social and economic analysis of integrated building transportation energy system

Zhengxuan Liu[1], Ying Sun[2] and Ruopeng Huang[3]
[1]Faculty of Architecture and the Built Environment, Delft University of Technology, Delft, The Netherlands
[2]School of Environmental and Municipal Engineering, Qingdao University of Technology, Qingdao, P.R. China
[3]School of Management Science and Real Estate, Chongqing University, Chongqing, P.R. China

13.1 Introduction

Numerous studies have demonstrated that global temperatures have climbed by 0.5°C−1°C over the past 100 years due to greenhouse gas emissions. The rise in global temperature has caused a series of environmental and ecological problems, including a dramatic increase in extreme weather [1−5]. Carbon emissions from the building and transportation sector account for a large proportion of total global carbon emissions. With rapid urbanization, energy consumption and carbon emissions from the building sector will also become more prominent [4,6−8]. Many countries have made significant efforts to tackle the increase in greenhouse gas emissions [9−11,12]. For instance, the Paris Agreement, an international treaty signed in 2016 under the United Nations Framework Convention on Climate Change, targets limiting global warming to well below preindustrial levels of 2°C and strives to limit temperature increases to 1.5°C. By 2021, 197 parties had ratified the Paris Agreement, which is regarded as a milestone achievement for global efforts to combat climate change. The Paris Agreement is widely recognized as the most ambitious and comprehensive international agreement on climate change to date [13−15]. To achieve this international agreement, one essential measure is to design buildings that consume less energy by adopting more renewable energy sources and using electric vehicles (EVs) on a large scale [16,17]. As representatives of decarbonization technologies, common renewable energy technologies include solar energy (photovoltaic and solar thermal) [18,19], geothermal energy (shallow geothermal [20−24], medium, and

Advances in Digitalization and Machine Learning for Integrated Building-Transportation Energy Systems
DOI: https://doi.org/10.1016/B978-0-443-13177-6.00006-0

deep geothermal), wind energy (wind power) [25,26], and also the applications of some advanced energy storage technologies (e.g., electrochemical and chemical energy storage [27−29], phase change energy storage [30−33]). Numerous studies have shown that CO_2 emissions from building and transportation sectors account for approximately 75% of total global carbon emissions [16,34]. Using cleaner electricity production, smart and flexible integration, and advanced energy management are necessary pathways and effective solutions to achieve carbon neutrality in buildings and transportation [35,36]. Therefore a mitigation strategy that combines the adoption of renewable energy for buildings with the use of fully EVs rather than traditional gasoline-powered vehicles appears to be most promising for achieving this target [37,38].

With the accelerated development of EV technology, the energy integration and synergy of the integrated building transportation energy system (IBTES) have attracted widespread attention from researchers, engineers, and manufacturers [39,40]. The synergistic function between buildings and EVs provides a favorable solution to increase the resilience and flexibility of the electric grid in the case of fluctuating energy supply from multiple energy systems. EVs can be used to support building energy demand through on-site charging and smart discharging of renewable energy systems, such as photovoltaics, thereby achieving energy sharing between EVs and buildings [41,42]. With the widespread deployment of EVs, the use of redundant storage capacity in EVs has been widely adopted to increase renewable energy penetration, mitigate renewable energy intermittency, and stabilize the grid [43,44]. In addition, the integration of EVs with buildings can improve the self-sufficiency of buildings' energy requirements and decrease their dependence on the power grid [45,46]. Multilateral synergistic functional benefits can be realized between buildings and transportation through their integrated applications.

In terms of technoeconomic aspects, it is currently important to consider the system performance matching and decarbonization capabilities of IBTES in different application scenarios. With the application of smart technologies, the monitoring and regulation capabilities are improved to make the management of IBTES more intelligent, thereby further increasing the application potential of this system [47−49]. For instance, Karan et al. [34] comparatively analyzed energy usage in buildings and transportation to provide an appropriate approach to assess the effectiveness of CO_2 mitigation strategies and the potential of CO_2 emission reductions. Studies showed that mitigation strategies, including EVs powered by conventional electricity, resulted in an average reduction of 3.7% in CO_2 emissions. The mitigation strategy using EVs obtained from grid-connected solar panels resulted in a 12.2% reduction in daily CO_2 emissions. Karan et al. [34] developed an appropriate approach to evaluate the potential CO_2 emission reductions and the effectiveness of greenhouse gas mitigation strategies in the building and transportation sector. Research showed that the average person produces about 20 lb of carbon dioxide per day, with 62% of that coming from transportation. EVs powered by electricity generated from coal-fired power plants can reduce CO_2 emissions by an average of 3.7%. EVs powered by solar energy obtained from grid-connected solar panels reduce CO_2 emissions by 12.2% per day (from 12.38 lb/day to 10.87 lb/day). This study also further demonstrated that combining EVs with off-grid power sources was the most successful strategy. It also estimated the initial cost per pound of CO_2 reduction per day for each mitigation strategy. Zhou et al. [50] systematically analyzed an application framework for energy integration and interaction between buildings and EVs containing different energy forms, advanced energy conversion, diverse energy storage systems, and hybrid grids. The buildings involved in this study include residential, public, and transportation buildings, and the involved EVs include conventional gasoline, biofuel, battery, and fuel cell EVs. Perspectives from the systematic review include different vehicle models, battery sizes, hydrogen tank capacities, and home charge and discharge rates. This study also systematically presented and discussed technical solutions for enhanced energy interactions from the perspective of buildings, such as on-site photovoltaics, on-site wind turbines, biofuels, and geothermal energy, and EVs, such as mobile storage and mobile renewable systems.

The integration of building energy systems with EVs has been seen as one of the main potential solutions of addressing greenhouse gas emissions in the building and transportation sectors [51−53]. As the social sciences are receiving increasing attention in energy studies, it has become more significant to explore IBTES from a social perspective. Therefore it is necessary to further study the social aspects to increase consumer awareness and acceptance of IBTES. This study provides an overview and analysis of recent research and application advancements in buildings, interactive EVs, and their interactions in facilitating IBTES to achieve energy flexibility and carbon footprint reduction from a social and economic perspective.

13.2 Social impacts and contributions of integrated building transportation energy system

After two decades of development, the innovation capacity of the socio-technical system of EVs has continuously increased, but it is still at a preliminary stage, which makes it difficult to meet users' requirements. Compared with the EV industry, IBTES is a more complex system, containing various interrelated elements that interact with each other. The advancement and implementation of IBTES in practice not only require technological developments but also the coordinated development of specific social factors. Therefore the analysis and summary of social aspects are particularly salient for the application and development of IBTES.

Kachirayil et al. [54] systematically reviewed 116 case studies of locally integrated energy system models to determine best practice approaches for model flexibility and addressing nontechnical constraints. There was rarely consideration of coupling with the transportation sector in the examples, especially EVs, although they could be used for smart charging or vehicle-to-grid (V2G) operations. And the societal aspects are often completely ignored. However, Rith et al. [55] demonstrated that improving community access to key services and facilities can promote equitable social development. Wu et al. [56] systematically summarized the positive contributions of the interactive trading behavior of buildings and EVs in establishing sustainable trading energy communities. An exploration of the physical space of energy, the cyberspace of data, and the social space of humans was presented. Low-carbon interactive energy solutions with key technologies and recent advances for net-zero energy buildings with high EV densities are discussed in a hierarchical manner.

In practice, many benefits of IBTES are inadequately understood, and social impacts are often overlooked in cost-benefit analyses. However, these benefits are closely related to social content and economic growth. Omahne et al. [57] systematically reviewed the research on the social impact of EVs. This study assesses the social impact of EVs by identifying the main current research priorities related to the perception of EVs. This study systematically divided the literature into the following social factors: acceptance, perception, impact, cost, welfare, and user experience. The findings indicated that existing studies are still lacking to assess the impacts of EVs on social well-being and user experience, but acceptance and user perceptions are frequently studied. In terms of potential user perceptions, the extended area of social perspectives of EVs is the research focus and trend. In addition, publications on the assessment of social impacts frequently study economic and environmental aspects in addition to social aspects. The results also indicated that 87.5% of the publications also integrated economic and environmental aspects when assessing the social impact of EVs. This is due to the fact that all aspects of sustainability are interconnected.

Wu et al. [58] adopted the socio-technical transition theory and multilevel perspective (MLP) approach in forecasting the transition route from conventional vehicles to new energy vehicles in the future. The study showed that the socio-technical system for new energy vehicles is still in its infancy and that many specific consumer requirements are difficult to satisfy. Fig. 13.1 illustrates the three MLP layers: ecological niche, regime, and landscape, for a four-stage transition. Both the supporting environment and the groundbreaking innovation of new energy vehicles have put pressure on the current sociotechnical system of the traditional vehicle industry. Meanwhile, the traditional vehicle system puts much pressure on the development of new energy vehicles, but the development of new energy vehicles is also supported by the exogenous environment. This study contributes to the formation of a sustainable low-carbon transformation route for the Chinese vehicle industry.

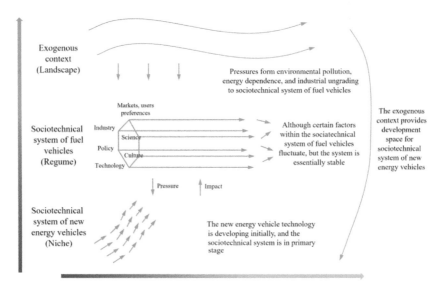

Figure 13.1 Interaction of the three MLP dimensions: landscape, system, and niche, in the sociotechnical system of the Chinese vehicle industry in four stages [58]. *MLP*, Multilevel perspective.

Dall-Orsoletta et al. [59] investigated possible related injustices in the life cycle of EVs. The findings clarified how EVs can contribute to flexibility justice through smart grid and V2G development. Fig. 13.2 represents the main aspects regarding distributive justice and flexibility justice based on life cycle stages being considered, along with positive (+), negative (−), or a combination of symbols (− +) indicating the potential impacts of EVs. As can be seen in Fig. 13.3, the distributive injustice is evenly distributed throughout the life cycle stages, which indicates the existence of potential negative impacts of EVs and is against the just energy transition. EVs seem to have the potential to positively impact flexibility justice, mainly in the distribution and operation stages. In addition, this study provides some suggestions for the development of the social aspects of EV technology to promote its inclusive social innovation.

V2G is an approach that improves the sustainability and reliability of the electric and transportation systems by enabling a two-way connection between them. The transition from conventional vehicles to V2G allows vehicles to simultaneously increase the efficiency (and profitability) of the grid, decrease greenhouse gas emissions from transportation, and save operational costs for vehicle owners and other users. To understand the state-of-the-art in this research area, Sovacool et al. [60] conducted a systematic review of 197 peer-reviewed articles that addressed V2G-related studies. The study revealed that there are still many social barriers to V2G

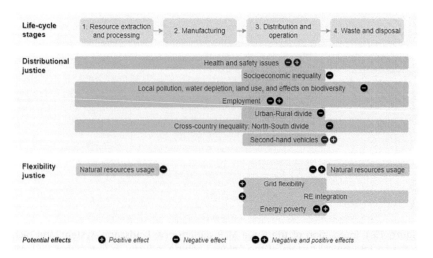

Figure 13.2 Distributive and flexible justice: the potential impacts of EVs [59].

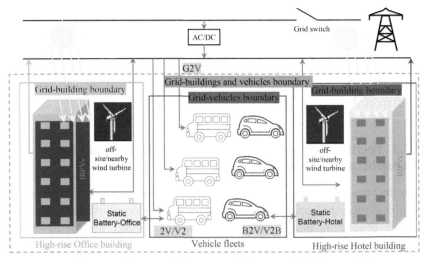

Figure 13.3 Schematic diagram of the renewable energy-buildings-vehicles energy-sharing network [42].

research, and many unexplored socially relevant questions remain for research. The complexities of business models, market segmentation, and the motivations of users have been virtually ignored in recent V2G studies. In addition, the study also clarified that the integrated energy communities need to consider a combination of human and social factors in their future studies.

The decentralized peer-to-peer energy-sharing technology is remarkably promising as the next-generation mechanism for smart building energy management that can facilitate the realization of near-net-zero energy buildings. In this background, Lyu et al. [61] proposed an integrated smart building energy-sharing framework that considers multiple dynamic components, including the heating, ventilation, and air conditioning system, the battery storage system, and EVs, aiming to maximize social benefits through peer-to-peer energy cooperation. An extensive case study based on a smart building community shows that the proposed peer-to-peer interaction framework can significantly improve the overall social welfare of the smart buildings involved.

In summary, with regard to government decision-making, the ministries and statistical agencies responsible for buildings and energy should collect data on sociocultural trends and more active community participation should be mandated to minimize communication and collaboration barriers [62—64]. The government could also recommend that socioculturalism be considered a core competency required by practices responsible for planning and implementing low-carbon transitions [65,66]. Policymakers should also pay more attention to the aspirations and capacities of groups and collective phenomena to shape and influence low-carbon transitions. Moreover, such socioculturally aware measures, as well as any resulting policies, will also need to be inclusive [67].

13.3 Technoeconomic analysis of integrated building transportation energy system

The techno-economic analysis of IBTES is an essential step to successfully implementing the application in practice, which contributes to optimizing energy utilization, informing decisions, reducing costs, improving energy security, decreasing carbon emissions, and achieving economic benefits [68−71]. A summary analysis of this integrated system from different perspectives can provide some valuable recommendations to decision-makers so that they can make informed decisions about IBTES implementation and ensure that the systems are implemented in a cost-effective and sustainable manner.

Plenty of studies have shown that by integrating EVs into building energy systems, the degradation cost of batteries will increase due to the increased charge/discharge cycles [72−74]. The IBTES is economically feasible only if the operational cost savings of the integrated system exceed the battery degradation cost. Therefore considering the battery depreciation cost, some researchers have developed a dynamic battery degradation model [42] and a fuel cell degradation model [16] that are applicable for IBTES. Based on the life-cycle economic performance analysis of this system, the battery performance was overestimated by 20% (from 2.404×10^8 to 3.005×10^8 HK$) if the battery cycle degradation was not taken into account [40].

Zhou et al. [42] proposed a resilient energy network with interactive renewable energy-buildings-vehicles energy sharing, as shown in Fig. 13.3. In this study, energy management is performed through a centralized collaborative controller of renewable energy and grid power to be responsive to the mobile consumption and energy demand of the buildings. The study also developed an advanced battery conservation energy control strategy for EVs to investigate equivalent CO_2 emissions, import costs, energy flexibility, and the equivalent relative capacity of battery storage. It also proposed a robust solution for relative capacity improvement by limiting the lower limit of the fractional state to 0.7. In addition, in another study by Zhou et al. [75], a transition framework from negative to positive regional energy-sharing networks, considering battery cycle degradation, advanced battery management strategies, and flexible building-vehicle interaction, was proposed, as shown in Fig. 13.4. The study also investigated the techno-economic performance, including net present value, discounted rate payback period, and net direct energy consumption. The results indicated that as the energy paradigm shifted from a negative to a positive system, the

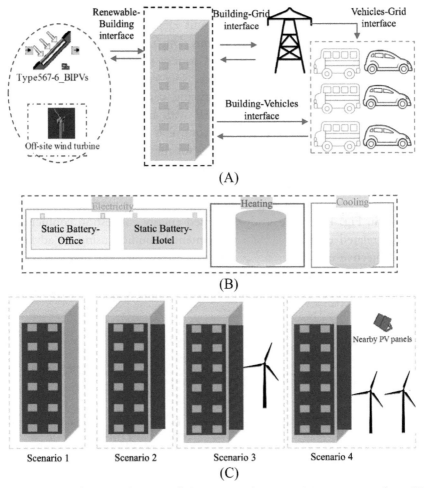

Figure 13.4 Schematic diagram of the proposed system: (A) system interface; (B) hybrid power and thermal energy storage; (C) energy paradigm transition from negative to positive system [75].

net present value increased from -7.182×10^7 to 5.164×10^8 HK\$ and the average annual net direct energy consumption decreased from 249.1 to $-343.3\,\mathrm{kWh/m^2}$. The study also demonstrated the techno-economic performance of the district building-vehicle system and the shifting of the energy paradigm from negative to positive and proposed a series of promising solutions.

With the advancement of smart homes and smart electric grids, buildings as the largest consumers of electricity can participate in this smart operation to realize the integrated application of building and EV interactive grid system. Mirakhorli et al. [76] presented a building energy management system that takes into account electricity prices and occupant behavior, as shown in Fig. 13.5. In this coupled management system, air conditioners, water heaters, EVs, and battery storage are controlled in a building equipped with photovoltaics. The results indicated that these methods of real-time five-minute pricing can achieve cost savings of 20%–30% on these devices compared with traditional control systems. When the added battery-optimized control is considered in the control strategy, the total electricity cost savings for these appliances is 42%.

Zhou et al. [77] developed an interactive building-vehicle energy-sharing network with multidirectional energy interactions as well as a grid response strategy for managing nonpeak renewables and grid power, as shown in Fig. 13.6. The study addressed energy congestion conflicts and energy-related economic and environmental conflicts using an advanced multiobjective optimization algorithm (i.e., the Pareto-archived NSGA-II algorithm) to achieve optimal design and reliable operations. Results showed that the formulated interactive building-vehicle energy-sharing

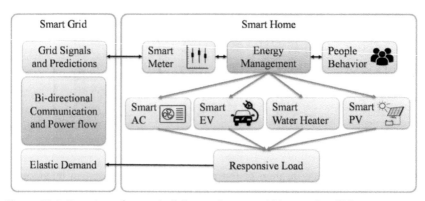

Figure 13.5 Overview of smart building and smart grid integration [76].

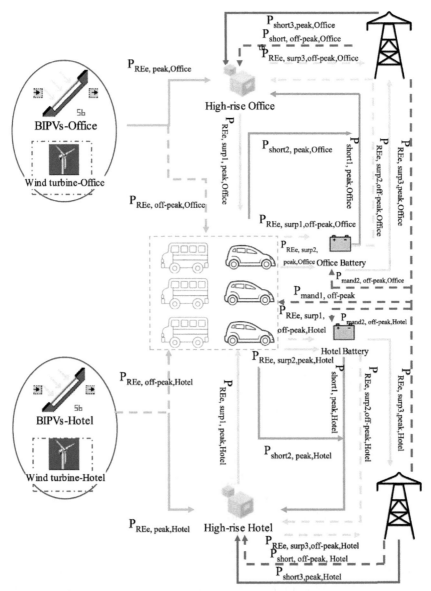

Figure 13.6 Schematic diagram of the interactive building-vehicle energy-sharing network [77].

network exhibits greater advantages over traditional isolated systems in terms of cost, emissions, and energy flexibility. Specifically, the optimized integrated system was able to decrease CO_2 emissions by 7.5%, from

147.4 to 136.4 kg/m$^2 \cdot$a, and import costs from the grid by 8.5%, from 212.7 to 194.6 HK\$/m$^2 \cdot$a, compared with the pre-optimized system. This study is critical for an interactive building–vehicle energy-sharing network with high-level energy flexibility for smart cities and can provide effective solutions and implementation recommendations for decision-makers.

13.4 Implementation challenges and recommendations of integrated building transportation energy system in social and economic aspects

Achieving ambitious greenhouse gas emissions targets will require multi-sectoral, participatory programs that address direct emissions as well as supply chain and end-user emissions [38,78]. In addition to the technical challenges, there will be dual challenges in the social and economic aspects. This section will discuss and analyze the implementation barriers to IBTES and the effective targeted solutions of these two aspects.

13.4.1 Social perspectives

Technological developments can effectively address some issues during the implementation of IBTES. However, technological developments alone are not sufficient to effectively promote the development of IBTES applications. It is equally important to focus on the social aspects, such as public awareness and acceptance, the participation of different stakeholders, public policies, and regulations.

13.4.1.1 Public perception and acceptance

Challenges: Public perception and understanding of the IBTES is critical to its successful implementation. Perceptions, for example, its benefits, costs, and impact on the environment, regarding the system need to be positive to gain public support and acceptance [79,80]. In practice, the general public may not have sufficient awareness of the concept and benefits of IBTES, resulting in a lack of acceptance and support. Implementation of this IBTES may require some major modifications to building and transportation infrastructure, including retrofitting buildings, upgrading transportation systems, and changing energy usage patterns [45,81]. As a new technology, the public may question its effectiveness and reliability, and users are concerned about the series of problems that will arise during its usage. The public may resist these changes, preferring to stick to familiar systems, even though the IBTES system may be more effective and sustainable.

Recommendations: To increase the acceptance of IBTES, the benefits and reliability of the system can be illustrated through demonstration projects, case studies, and actual monitored data. It is necessary to share the monitored data and results of these projects with the public, including performance metrics and realized savings. Public awareness and understanding of this integrated system and its benefits can also be improved through the development and implementation of educational campaigns, workshops, and other outreach activities. In addition, regular monitoring and evaluation of the impacts of this integrated system on the public and the environment are necessary to make adjustments and improve the system over time. These measures include establishing feedback mechanisms, such as surveys and focus groups, as well as performance metrics and data analysis. Based on this, addressing concerns and fears about the applications of these new technologies through open and transparent communication will also help build public support for their usage.

13.4.1.2 *Cooperation and participation of stakeholders*

Challenges: The collaboration and engagement of stakeholders are critical aspects of the successful implementation of some advanced energy systems [82–84]. Therefore when promoting the implementation of IBTES in practice, different stakeholders may have different priorities and conflicting interests among them, which can hinder cooperation and engagement due to a lack of trust. On the other hand, different stakeholders may have various levels of expertise and experience, and they may have different viewpoints on promoting the implementation of IBTES, which may lead to misunderstandings and communication barriers to some extent. Stakeholders may also oppose the changes due to a lack of understanding of the advantages and benefits of IBTES. In addition, it is difficult for many stakeholders to participate in decision-making about the implementation of IBTES, which can affect trust and prevent collaboration among different stakeholders.

Recommendations: Creating trust among stakeholders is essential in ensuring their effective collaboration and participation, which can be achieved through open and transparent communication, as well as creating opportunities for feedback and input from stakeholders [85,86]. Effective communication and knowledge sharing are essential in ensuring that all stakeholders have a mutual understanding of the integrated system and its benefits. This is accomplished through regular training and education programs and opportunities for knowledge sharing and collaboration. Encouraging the participation of stakeholders in the decision-making process contributes to building trust and fostering collaboration, which can be achieved through regular meetings and joint decision-making processes between different stakeholders.

13.4.1.3 Incentive measures for participants

Challenges: Numerous studies have shown that the process of the practical application of new energy systems usually lacks effective incentives, which decreases the usage rate of the technologies [87,88]. Conventional energy plants and energy efficiency service companies do not prefer to adopt a decentralized model to replace their already well-established revenue streams. Therefore targeted incentives are necessary to promote the participation of all energy stakeholders in the implementation of IBTES, such as price incentives that allow consumers to choose a nearby distributed energy retailer, and policy incentives that allow utilities to coordinate and distribute peer-to-peer energy transactions in the district community.

Recommendations: For this integrated energy system incentives must be developed to satisfy the diverse requirements of users with different roles and functions in the decentralized environment. Through decentralized identity management, the integration and extension of various roles, responsibilities, and interactions of decentralized market actors at different levels can be supported. The customized incentives must be deployed in different organizations and individuals, which will encourage an increasing number of participants and facilitate a more interoperable, consistent, transparent, and trustworthy energy society.

13.4.1.4 Regulatory policies and government support

Challenges: Regulatory policies and restrictions can pose significant barriers to IBTES implementation [89]. The inconsistency of regulations and related policies in different jurisdictions can cause many problems in the implementation of IBTES. In practice, there is still a lack of standard procedures for implementing IBTES, making its implementation less efficient and the whole process more time-consuming and complex. In addition, national and local government support is critical to the successful implementation of IBTES, as regulatory policies and restrictions are typically driven by government decision-making [89,90].

Recommendations: To overcome the barriers of inconsistent regulations, it may be necessary to develop consistent common standards and regulations for the system across regions and to harmonize policies to minimize confusion and improve predictability. To mitigate the barriers of standardized procedures, it may be necessary to develop common protocols and procedures for different regions, and simplify and streamline the permitting process, and make the application process more straightforward and user-friendly to ensure the smooth implementation and effective operation of the integrated system. In addition, it is beneficial to build political support for the implementation and advancement of this system by engaging different stakeholders and achieving consensus on their interests and potential impacts.

Overall, to overcome these societal barriers, a comprehensive and coordinated approach that involves all relevant stakeholders, including government agencies, building owners and suppliers, and the public, will be required in future studies. This can ensure that regulations and policies are introduced that effectively support the implementation of the system and avoid creating unintended barriers to its implementation. Effective stakeholder engagement and collaboration can contribute to building an understanding of this system and identifying and addressing some issues that arise during the advancement process.

13.4.2 Economic perspectives

The implementation challenges and recommendations of IBTES from the economic perspective include the initial build cost of the system and the maintenance costs, uncertain return-on-investment cycle, financial incentive measures, and market demand and competition of the system.

13.4.2.1 System initial build and maintenance costs

Challenges: The implementation of IBTES requires a significant initial investment, which may include the cost of design and construction, as well as retrofitting existing buildings and transportation systems. The high up-front costs can make it difficult for building owners, energy providers, and transportation operators to adopt the system, especially in areas with limited financial resources or incentives. The major factors hindering the development of the IBTES market also include high equipment purchase prices, and high vehicle depreciation costs [91]. Many users are unsure about the cost of charging their cars or exactly how much power is required. This lack of knowledge hinders the calculation of possible costs and savings and therefore leads to inaccurate cost comparisons [92]. The battery lifetime of EVs is generally limited, and it may be a concern for many users that the costs of battery replacement will outweigh the lower operating costs, resulting in no significant economic benefit from EVs [93]. In addition, upgrading existing buildings and transportation infrastructure to support an IBTES may be expensive. These costs include the costs of equipment and technologies, such as smart energy management systems, EV charging stations, and building automation systems.

Recommendations: Faced with the high initial build and maintenance cost of IBTES, a number of scientifically reasonable measures need to be taken to overcome them. For example, obtaining funding and grants from government agencies, private sector organizations, and other sources can contribute to offsetting the high costs of technology and infrastructure upgrades. Leveraging existing building and transportation infrastructure, such as existing electrical and communications systems, can help reduce the costs of renovation and upgrading. The implementation of cost-effective technologies, such as low-cost building automation systems and EV charging stations, is available to assist in reducing costs and making integrated systems more affordable.

13.4.2.2 Uncertain return-on-investment cycle

Challenge: Users may be hesitant to invest in IBTES due to concerns about return on investment (ROI). This uncertainty stems from multiple factors, including the lack of standard metrics to assess ROI, the complexity of technologies, and limited case studies. The lack of standard metrics to assess the ROI of IBTES hinders the process of quantifying the benefits and accurately calculating the ROI [94,95]. Lack of standardization can cause confusion and mistrust among stakeholders and complicate the procedure of securing funding and support for IBTES implementation. In addition, technical complexity makes it challenging to quantify energy savings and benefits and increases uncertainty about the ROI.

Recommendations: The uncertainty of the ROI cycle can make it difficult to obtain funding and support for IBTES, especially when there is limited data and experience with these systems. The development of standard metrics to assess the ROI for IBTES can help quantify the benefits, accurately calculate the ROI, reduce uncertainty, and make it easier to obtain funding and system support. Encouraging collaboration among stakeholders, including building and transportation providers and technology providers, can help address the technical complexities of IBTES and increase the availability of data and case studies on the ROI of the integrated systems.

13.4.2.3 Financial incentive measures

Challenges: One of the main barriers to implementing financial incentives is the lack of awareness among users about the existence of these incentives and the potential benefits they can bring [96−98]. Many users may not be sufficiently aware of the policies and opportunities that exist for financial support. In addition, many users may perceive the application process for financial incentives as potentially complex and time-consuming because the process can involve extensive documentation and verification of information. This can be a challenge for users who are unfamiliar with the process, discouraging them from pursuing these financial incentives. The availability of financial incentives may be limited, which creates some uncertainty for users who are considering investing in the integrated system. The restrictions on the availability of financial resources can make it challenging for some users to obtain effective incentives [89,99].

Recommendations: To overcome these barriers, a number of measures can be adopted for the promotion of financial incentives for IBTES. These measures include awareness campaigns, seminars, and training sessions to increase users' understanding of the system's benefits and the incentives that support them [100,101]. To minimize the burden of applying for incentives, it is possible to simplify the application process. This can be achieved by streamlining the documentation and verification processes and providing clear and comprehensive guidance on the application process.

13.4.2.4 Market demand and competition

Challenges: The success of IBTES is dependent on the level of market demand, which can be influenced by several factors, including energy costs, public awareness, and availability. The demand level for IBTES can also be influenced by the level of competition from other energy systems. This is because users may not be aware of the benefits of these systems, or may not understand their potential benefits. The market demand for IBTES remains low at this stage, especially in areas where people are not yet aware of the benefits of such systems. For these systems to be successful, there needs to be a significant level of interest and demand from users. In addition, the lack of public awareness and understanding of these systems may limit their ability to compete with more mature energy sources, such as geothermal energy, which may also make it difficult for these new energy systems to establish a firm foothold in the marketplace [102,103].

Recommendations: To overcome these barriers, it is necessary to emphasize the building of awareness among users. It is critical to provide education about the potential benefits of these systems so that users can realize the potential savings on their energy bills and the positive impact of these systems on the environment. It is important to search for solutions to reduce component costs, optimize energy production, and improve energy efficiency. In addition, partnerships with other organizations and businesses can help increase market demands and contribute to decreasing competition by sharing resources and knowledge. Government incentives can play a key role in increasing market demands for new energy systems, including tax credits, subsidies, and grants, which can be used to make these systems more affordable and accessible, thereby increasing market demands and diminishing competition from conventional energy sources [104–106].

13.5 Conclusions and future studies

This chapter summarizes the current application and development status of IBTES from two different perspectives, social and economic. The results illustrated that IBTES has a large decarbonization potential and can effectively mitigate carbon emissions in the building and transportation sector. The system also has promising social and economic benefits as well as good potential for applications. In addition, this chapter also discusses the main issues and challenges in the practical application of IBTES from social and economic perspectives, including low social acceptance; unfamiliarity with the technology; poor stakeholder cooperation, participation, and motivation; high initial build and maintenance costs of the system; and inadequate incentives. Based on these existing challenges, this chapter proposes recommendations to effectively promote the implementation of IBTES in practice. This study can provide theoretical guidance and suggestions in IBETS-related industrial developments for policymakers.

As the role of IBTES becomes increasingly important in achieving global carbon neutrality, future studies will be conducted in the following areas to improve their practical applications:

1. An analysis of current and future market trends around IBTES can be conducted in the future. This includes consumer demand, adoption rates, and market growth, as well as the key driving factors for market demand and competition, which will provide valuable insights into the most effective strategies for increasing market demand and reducing competition, as well as opportunities for growth and improvement.

2. Future studies are expected to integrate the user experience of IBTES with the evaluation of social welfare. In addition, the evaluation of the social aspects of IBTES should incorporate the assessment of perception or acceptance to facilitate the commercialization of the system.

3. Comparative studies of IBTES in different regions and countries can be conducted in the future, including its implementation, adoption rates, and market demands. This will help identify best practices and lessons learned from other regions, as well as opportunities for their improvement.

4. A life-cycle analysis of IBTES should be considered, including the environmental and social impacts, as well as their economic costs and benefits. This will help identify the most sustainable and cost-effective strategies for implementing these systems, as well as the overall environmental and social impacts.

References

[1] Mikhaylov A, Moiseev N, Aleshin K, Burkhardt T. Global climate change and greenhouse effect. Entrepreneurship and Sustainability Issues 2020;7:2897.

[2] Park M, Wang Z, Li L, Wang X. Multi-objective building energy system optimization considering EV infrastructure. Applied Energy 2023;332:120504.

[3] Sun Y, Haghighat F, Fung BCM. A review of the-state-of-the-art in data-driven approaches for building energy prediction. Energy and Buildings 2020;221:110022.

[4] Sun Y, Panchabikesan K, Haghighat F, Luo J, Moreau A, Robichaud M. Development of advanced controllers to extend the peak shifting possibilities in the residential buildings. Journal of Building Engineering 2021;43:103026.

[5] Zhou Y, Zheng S, Liu Z, Wen T, Ding Z, Yan J, et al. Passive and active phase change materials integrated building energy systems with advanced machine-learning based climate-adaptive designs, intelligent operations, uncertainty-based analysis and optimisations: A state-of-the-art review. Renewable and Sustainable Energy Reviews 2020;130:109889.

[6] He Y, Zhou Y, Liu J, Liu Z, Zhang G. An inter-city energy migration framework for regional energy balance through daily commuting fuel-cell vehicles. Applied Energy 2022;324:119714.

[7]   Liu Z, Xie M, Zhou Y, He Y, Zhang L, Zhang G, et al. A state-of-the-art review on shallow geothermal ventilation systems with thermal performance enhancement system classifications, advanced technologies and applications. Energy and Built Environment 2023;4:148−68.

[8]   Liu Z, Yu Z, Yang T, Qin D, Li S, Zhang G, et al. A review on macro-encapsulated phase change material for building envelope applications. Building and Environment 2018;144:281−94.

[9]   Guillén-Lambea S, Carvalho M. A critical review of the greenhouse gas emissions associated with parabolic trough concentrating solar power plants. Journal of Cleaner Production 2021;289:125774.

[10] Heidari N, Pearce JM. A review of greenhouse gas emission liabilities as the value of renewable energy for mitigating lawsuits for climate change related damages. Renewable and Sustainable Energy Reviews 2016;55:899−908.

[11] Sun Y, Panchabikesan K, Joybari MM, Olsthoorn D, Moreau A, Robichaud M, et al. Enhancement in peak shifting and shaving potential of electrically heated floor residential buildings using heat extraction system. Journal of Energy Storage 2018;18:435−46.

[12] Liu Z, Zhang X, Sun Y, Zhou Y. Advanced controls on energy reliability, flexibility and occupant-centric control for smart and energy-efficient buildings. Energy and Buildings 2023;113436. Available from: https://doi.org/10.1016/j.enbuild.2023.113436.

[13] Aleluia Reis L, Tavoni M. Glasgow to Paris—the impact of the Glasgow commitments for the Paris climate agreement. iScience 2023;26:105933.

[14] You K, Yu Y, Cai W, Liu Z. The change in temporal trend and spatial distribution of CO_2 emissions of China's public and commercial buildings. Building and Environment 2023;229:109956.

[15] Zhou Y, Liu Z. A cross-scale 'material-component-system' framework for transition towards zero-carbon buildings and districts with low, medium and high-temperature phase change materials. Sustainable Cities and Society 2023;89:104378.

[16] He Y, Zhou Y, Wang Z, Liu J, Liu Z, Zhang G. Quantification on fuel cell degradation and techno-economic analysis of a hydrogen-based grid-interactive residential energy sharing network with fuel-cell-powered vehicles. Applied Energy 2021;303:117444.

[17] Liu Z, Sun Y, Xing C, Liu J, He Y, Zhou Y, et al. Artificial intelligence powered large-scale renewable integrations in multi-energy systems for carbon neutrality transition: challenges and future perspectives. Energy and AI 2022;10:100195.

[18] Zhou Y, Liu X, Zhang G. Performance of buildings integrated with a photo-voltaic−thermal collector and phase change materials. Procedia Engineering 2017;205:1337−43.

[19] Zhou Y, Zheng S. Multi-level uncertainty optimisation on phase change materials integrated renewable systems with hybrid ventilations and active cooling. Energy 2020;202:117747.

[20] Liu J, Yu Z, Liu Z, Qin D, Zhou J, Zhang G. Performance analysis of earth-air heat exchangers in hot summer and cold winter areas. Procedia Engineering 2017;205:1672−7.

[21] Liu Z, Qian QK, Visscher H, Zhang G. Review on shallow geothermal promoting energy efficiency of existing buildings in Europe. IOP Conference Series: Earth and Environmental Science. IOP Publishing; 2022, p. 012026.

[22] Liu, Z., Roccamena, L., Mankibi, M.E., Yu, Z.J., Yang, T., Sun, Y., et al., 2018a. Experimental analysis and model verification of a new earth-to-air heat exchanger system, 2018 4th International Conference on Renewable Energies for Developing Countries (REDEC), pp. 1−7.

[23] Liu Z, Yu Z, Yang T, Li S, El Mankibi M, Roccamena L, et al. Experimental investigation of a vertical earth-to-air heat exchanger system. Energy Conversion and Management 2019;183:241−51.

[24] Tang, L., Liu, Z., Zhou, Y., Qin, D., Zhang, G., 2020. Study on a dynamic numerical model of an underground air tunnel system for cooling applications—experimental validation and multidimensional parametrical analysis, energies.

[25] Desalegn B, Gebeyehu D, Tamirat B. Wind energy conversion technologies and engineering approaches to enhancing wind power generation: a review. Heliyon 2022;8:e11263.

[26] Zheng X, He L, Wang S, Liu X, Liu R, Cheng G. A review of piezoelectric energy harvesters for harvesting wind energy. Sensors and Actuators A: Physical 2023;352:114190.

[27] Liu J, Zhou Y, Yang H, Wu H. Net-zero energy management and optimization of commercial building sectors with hybrid renewable energy systems integrated with energy storage of pumped hydro and hydrogen taxis. Applied Energy 2022;321:119312.

[28] Zhou L, Zhou Y. Study on thermo-electric-hydrogen conversion mechanisms and synergistic operation on hydrogen fuel cell and electrochemical battery in energy flexible buildings. Energy Conversion and Management 2023;277:116610.

[29] Zhou Y. A dynamic self-learning grid-responsive strategy for battery sharing economy—multi-objective optimisation and posteriori multi-criteria decision making. Energy 2023;266:126397.

[30] Liu Z, Sun P, Xie M, Zhou Y, He Y, Zhang G, et al. Multivariant optimization and sensitivity analysis of an experimental vertical earth-to-air heat exchanger system integrating phase change material with Taguchi method. Renewable Energy 2021;173:401−14.

[31] Liu Z, Yu Z, Yang T, El Mankibi M, Roccamena L, Sun Y, et al. Experimental and numerical study of a vertical earth-to-air heat exchanger system integrated with annular phase change material. Energy Conversion and Management 2019;186:433−49.

[32] Qin D, Liu Z, Zhou Y, Yan Z, Chen D, Zhang G. Dynamic performance of a novel air-soil heat exchanger coupling with diversified energy storage components—modelling development, experimental verification, parametrical design and robust operation. Renewable Energy 2021;167:542−57.

[33] Zhou Y, Liu Z, Zheng S. 15 − Influence of novel PCM-based strategies on building cooling performance. In: Pacheco-Torgal F, Czarnecki L, Pisello AL, Cabeza LF, Granqvist C-G, editors. Eco-efficient materials for reducing cooling needs in buildings and construction. Woodhead Publishing; 2021, p. 329−53.

[34] Karan E, Mohammadpour A, Asadi S. Integrating building and transportation energy use to design a comprehensive greenhouse gas mitigation strategy. Applied Energy 2016;165:234−43.

[35] Liu J, Yang H, Zhou Y. Peer-to-peer energy trading of net-zero energy communities with renewable energy systems integrating hydrogen vehicle storage. Applied Energy 2021;298:117206.

[36] Zhou Y. Transition towards carbon-neutral districts based on storage techniques and spatiotemporal energy sharing with electrification and hydrogenation. Renewable and Sustainable Energy Reviews 2022;162:112444.

[37] Liu Z, Zhou Y, Yan J, Marcos T. Frontier ocean thermal/power and solar PV systems for transformation towards net-zero communities. Energy 2023;284:128362. Available from: https://doi.org/10.1016/j.energy.2023.128362.

[38] You K, Yu Y, Cai W, Liu Z. The change in temporal trend and spatial distribution of CO_2 emissions of China's public and commercial buildings. Building and Environment 2023;229:109956. Available from: https://doi.org/10.1016/j.buildenv.2022.109956.

[39] Zhou Y, Cao S. Energy flexibility investigation of advanced grid-responsive energy control strategies with the static battery and electric vehicles: a case study of a high-rise office building in Hong Kong. Energy Conversion and Management 2019;199:111888.

[40] Zhou Y, Cao S. Coordinated multi-criteria framework for cycling aging-based battery storage management strategies for positive building−vehicle system with

renewable depreciation: life-cycle based techno-economic feasibility study. Energy Conversion and Management 2020;226:113473.

[41] Zhou Y. Climate change adaptation with energy resilience in energy districts—a state-of-the-art review. Energy and Buildings 2023;279:112649.

[42] Zhou Y, Cao S, Hensen JLM, Hasan A. Heuristic battery-protective strategy for energy management of an interactive renewables—buildings—vehicles energy sharing network with high energy flexibility. Energy Conversion and Management 2020;214:112891.

[43] He Y, Zhou Y, Yuan J, Liu Z, Wang Z, Zhang G. Transformation towards a carbon-neutral residential community with hydrogen economy and advanced energy management strategies. Energy Conversion and Management 2021;249:114834.

[44] Huang P, Munkhammar J, Fachrizal R, Lovati M, Zhang X, Sun Y. Comparative studies of EV fleet smart charging approaches for demand response in solar-powered building communities. Sustainable Cities and Society 2022;85:104094.

[45] Nematchoua MK, Marie-Reine Nishimwe A, Reiter S. Towards nearly zero-energy residential neighbourhoods in the European Union: a case study. Renewable and Sustainable Energy Reviews 2021;135:110198.

[46] Zhou Y. Artificial intelligence in renewable systems for transformation towards intelligent buildings. Energy and AI 2022;10:100182.

[47] Heredia WB, Chaudhari K, Meintz A, Jun M, Pless S. Evaluation of smart charging for electric vehicle-to-building integration: a case study. Applied Energy 2020;266:114803.

[48] Wang Z, Wang L, Dounis AI, Yang R. Integration of plug-in hybrid electric vehicles into energy and comfort management for smart building. Energy and Buildings 2012;47:260−6.

[49] Wu X, Hu X, Teng Y, Qian S, Cheng R. Optimal integration of a hybrid solar-battery power source into smart home nanogrid with plug-in electric vehicle. Journal of Power Sources 2017;363:277−83.

[50] Zhou Y, Cao S, Hensen JLM, Lund PD. Energy integration and interaction between buildings and vehicles: a state-of-the-art review. Renewable and Sustainable Energy Reviews 2019;114:109337.

[51] Pearre NS, Ribberink H. Review of research on V2X technologies, strategies, and operations. Renewable and Sustainable Energy Reviews 2019;105:61−70.

[52] Sadeghian O, Moradzadeh A, Mohammadi-Ivatloo B, Abapour M, Anvari-Moghaddam A, Shiun Lim J, et al. A comprehensive review on energy saving options and saving potential in low voltage electricity distribution networks: building and public lighting. Sustainable Cities and Society 2021;72:103064.

[53] Sadeghian O, Oshnoei A, Mohammadi-ivatloo B, Vahidinasab V, Anvari-Moghaddam A. A comprehensive review on electric vehicles smart charging: solutions, strategies, technologies, and challenges. Journal of Energy Storage 2022;54:105241.

[54] Kachirayil F, Weinand JM, Scheller F, McKenna R. Reviewing local and integrated energy system models: insights into flexibility and robustness challenges. Applied Energy 2022;324:119666.

[55] Rith M, Roquel KIDZ, Lopez NSA, Fillone AM, Biona JBMM. Towards more sustainable transport in Metro Manila: a case study of household vehicle ownership and energy consumption. Transportation Research Interdisciplinary Perspectives 2020;6:100163.

[56] Wu Y, Wu Y, Guerrero JM, Vasquez JC. Decentralized transactive energy community in edge grid with positive buildings and interactive electric vehicles. International Journal of Electrical Power & Energy Systems 2022;135:107510.

[57] Omahne V, Knez M, Obrecht M. Social aspects of electric vehicles research—trends and relations to sustainable development goals. World Electric Vehicle Journal 2021.

[58] Wu Z, Shao Q, Su Y, Zhang D. A socio-technical transition path for new energy vehicles in China: a multi-level perspective. Technological Forecasting and Social Change 2021;172:121007.

[59] Dall-Orsoletta A, Ferreira P, Gilson Dranka G. Low-carbon technologies and just energy transition: prospects for electric vehicles. Energy Conversion and Management: X 2022;16:100271.

[60] Sovacool BK, Noel L, Axsen J, Kempton W. The neglected social dimensions to a vehicle-to-grid (V2G) transition: a critical and systematic review. Environmental Research Letters 2018;13:013001.

[61] Lyu C, Jia Y, Xu Z. Fully decentralized peer-to-peer energy sharing framework for smart buildings with local battery system and aggregated electric vehicles. Applied Energy 2021;299:117243.

[62] Guo Y, Zhang Z, Burçin B, Lin Q. Novel issues for urban energy-saving management: renewal of leftover space. Sustainable Energy Technologies and Assessments 2023;55:102934.

[63] Meelen T, Truffer B, Schwanen T. Virtual user communities contributing to upscaling innovations in transitions: the case of electric vehicles. Environmental Innovation and Societal Transitions 2019;31:96−109.

[64] Xinyue Fu, Queena Qian, Guiwen Liu, Taozhi Zhuang, Henk Visscher, Ruopeng Huang. Overcoming inertia for sustainable urban development: Understanding the role of stimuli in shaping residents' participation behaviors in neighborhood regeneration projects in China. Environmental Impact Assessment Review 2023;103:107252.

[65] Huang R, Liu G, Li K, Liu Z, Fu X, Wen J. Evolution of residents' cooperative behavior in neighborhood renewal: An agent-based computational approach. Computers, Environment and Urban Systems 2023;105:102022.

[66] Shima E, Queena K, Gerdien de V, Henk V. Municipal governance and energy retrofitting of owner-occupied homes in the Netherlands. Energy and Buildings 2022;274:112423.

[67] Al-Zo'by M. Culture and the politics of sustainable development in the GCC: identity between heritage and globalization. Development in Practice 2019;29:559−69.

[68] Huang P, Lovati M, Zhang X, Bales C, Hallbeck S, Becker A, et al. Transforming a residential building cluster into electricity prosumers in Sweden: Optimal design of a coupled PV-heat pump-thermal storage-electric vehicle system. Applied Energy 2019;255:113864.

[69] Zheng S, Huang G, Lai ACK. Coordinated energy management for commercial prosumers integrated with distributed stationary storages and EV fleets. Energy and Buildings 2023;282:112773.

[70] Zhou Y. Energy sharing and trading on a novel spatiotemporal energy network in Guangdong-Hong Kong-Macao Greater Bay Area. Applied Energy 2022;318:119131.

[71] Chang S, Cho J, Heo J, Kang J, Kobashi T. Energy infrastructure transitions with PV and EV combined systems using techno-economic analyses for decarbonization in cities. Applied Energy 2022;319:119254.

[72] Ekhteraei Toosi H, Merabet A, Swingler A. Impact of battery degradation on energy cost and carbon footprint of smart homes. Electric Power Systems Research 2022;209:107955.

[73] Huang P, Tu R, Zhang X, Han M, Sun Y, Hussain SA, et al. Investigation of electric vehicle smart charging characteristics on the power regulation performance in solar powered building communities and battery degradation in Sweden. Journal of Energy Storage 2022;56:105907.

[74] Ouédraogo S, Faggianelli GA, Pigelet G, Notton G, Duchaud JL. Performances of energy management strategies for a Photovoltaic/Battery microgrid considering battery degradation. Solar Energy 2021;230:654−65.

[75] Zhou Y, Cao S, Hensen JLM. An energy paradigm transition framework from negative towards positive district energy sharing networks—battery cycling aging, advanced battery management strategies, flexible vehicles-to-buildings interactions, uncertainty and sensitivity analysis. Applied Energy 2021;288:116606.

[76] Mirakhorli A, Dong B. Market and behavior driven predictive energy management for residential buildings. Sustainable Cities and Society 2018;38:723−35.

[77] Zhou Y, Cao S, Kosonen R, Hamdy M. Multi-objective optimisation of an interactive buildings-vehicles energy sharing network with high energy flexibility using the Pareto archive NSGA-II algorithm. Energy Conversion and Management 2020;218:113017.

[78] Liu Z, Yu C, Qian Q, Huang R, You K, Henk V, et al. Incentive initiatives on energy-efficient renovation of existing buildings towards carbon−neutral blueprints in China: Advancements, challenges and prospects. Energy and Buildings 2023;296:113343. Available from: https://doi.org/10.1016/j.enbuild.2023.113343.

[79] López-Lambas ME, Monzón A, Pieren G. Analysis of using electric car for urban mobility, perceived satisfaction among university users. Transportation Research Procedia 2017;27:524−30.

[80] Brown MA, Soni A. Expert perceptions of enhancing grid resilience with electric vehicles in the United States. Energy Research & Social Science 2019;57:101241.

[81] Ahmad A, Khan JY. Optimal sizing and management of distributed energy resources in smart buildings. Energy 2022;244:123110.

[82] Hoarau Q, Perez Y. Interactions between electric mobility and photovoltaic generation: a review. Renewable and Sustainable Energy Reviews 2018;94:510−22.

[83] Lopez-Behar D, Tran M, Mayaud JR, Froese T, Herrera OE, Merida W. Putting electric vehicles on the map: a policy agenda for residential charging infrastructure in Canada. Energy Research & Social Science 2019;50:29−37.

[84] Warth J, von der Gracht HA, Darkow I-L. A dissent-based approach for multi-stakeholder scenario development—the future of electric drive vehicles. Technological Forecasting and Social Change 2013;80:566−83.

[85] Bakker S, Maat K, van Wee B. Stakeholders interests, expectations, and strategies regarding the development and implementation of electric vehicles: the case of the Netherlands. Transportation Research Part A: Policy and Practice 2014;66:52−64.

[86] Sopjani L, Stier JJ, Ritzén S, Hesselgren M, Georén P. Involving users and user roles in the transition to sustainable mobility systems: the case of light electric vehicle sharing in Sweden. Transportation Research Part D: Transport and Environment 2019;71:207−21.

[87] Saber H, Ranjbar H, Ehsan M, Anvari-Moghaddam A. Transactive charging management of electric vehicles in office buildings: a distributionally robust chance-constrained approach. Sustainable Cities and Society 2022;87:104171.

[88] Zou W, Sun Y, Gao D-C, Zhang X, Liu J. A review on integration of surging plug-in electric vehicles charging in energy-flexible buildings: Impacts analysis, collaborative management technologies, and future perspective. Applied Energy 2023;331:120393.

[89] Lopez-Behar D, Tran M, Froese T, Mayaud JR, Herrera OE, Merida W. Charging infrastructure for electric vehicles in Multi-Unit Residential Buildings: mapping feedbacks and policy recommendations. Energy Policy 2019;126:444−51.

[90] Thompson AW, Perez Y. Vehicle-to-Everything (V2X) energy services, value streams, and regulatory policy implications. Energy Policy 2020;137:111136.

[91] Axsen J, Orlebar C, Skippon S. Social influence and consumer preference formation for pro-environmental technology: the case of a U.K. workplace electric-vehicle study. Ecological Economics 2013;95:96−107.

[92] Graham-Rowe E, Gardner B, Abraham C, Skippon S, Dittmar H, Hutchins R, et al. Mainstream consumers driving plug-in battery-electric and plug-in hybrid electric cars: a qualitative analysis of responses and evaluations. Transportation Research Part A: Policy and Practice 2012;46:140−53.

[93] Daziano RA, Chiew E. Electric vehicles rising from the dead: data needs for forecasting consumer response toward sustainable energy sources in personal transportation. Energy Policy 2012;51:876−94.

[94] Gough R, Dickerson C, Rowley P, Walsh C. Vehicle-to-grid feasibility: a techno-economic analysis of EV-based energy storage. Applied Energy 2017;192:12−23.

[95] Sun C, Zhao X, Qi B, Xiao W, Zhang H. Economic and environmental analysis of coupled PV-energy storage-charging station considering location and scale. Applied Energy 2022;328:119680.

[96] Clinton BC, Steinberg DC. Providing the Spark: impact of financial incentives on battery electric vehicle adoption. Journal of Environmental Economics and Management 2019;98:102255.

[97] Hardman S, Chandan A, Tal G, Turrentine T. The effectiveness of financial purchase incentives for battery electric vehicles − a review of the evidence. Renewable and Sustainable Energy Reviews 2017;80:1100−11.

[98] Münzel C, Plötz P, Sprei F, Gnann T. How large is the effect of financial incentives on electric vehicle sales? − A global review and European analysis. Energy Economics 2019;84:104493.

[99] Tarei PK, Chand P, Gupta H. Barriers to the adoption of electric vehicles: evidence from India. Journal of Cleaner Production 2021;291:125847.

[100] Sierzchula W, Bakker S, Maat K, van Wee B. The influence of financial incentives and other socio-economic factors on electric vehicle adoption. Energy Policy 2014;68:183−94.

[101] Wu YA, Ng AW, Yu Z, Huang J, Meng K, Dong ZY. A review of evolutionary policy incentives for sustainable development of electric vehicles in China: strategic implications. Energy Policy 2021;148:111983.

[102] Jensen AF, Thorhauge M, Mabit SE, Rich J. Demand for plug-in electric vehicles across segments in the future vehicle market. Transportation Research Part D: Transport and Environment 2021;98:102976.

[103] Brand C, Cluzel C, Anable J. Modeling the uptake of plug-in vehicles in a heterogeneous car market using a consumer segmentation approach. Transportation Research Part A: Policy and Practice 2017;97:121−36.

[104] Gong S, Ardeshiri A, Hossein Rashidi T. Impact of government incentives on the market penetration of electric vehicles in Australia. Transportation Research Part D: Transport and Environment 2020;83:102353.

[105] Baumgarte F, Kaiser M, Keller R. Policy support measures for widespread expansion of fast charging infrastructure for electric vehicles. Energy Policy 2021;156:112372.

[106] Yang Z, Li Q, Yan Y, Shang W-L, Ochieng W. Examining influence factors of Chinese electric vehicle market demand based on online reviews under moderating effect of subsidy policy. Applied Energy 2022;326:120019.

Index

Note: Page numbers followed by "*f*" and "*t*" refer to figures and tables, respectively.